Television and the Modernization Ideal in 1980s China

Television and the Modernization Ideal in 1980s China

Dazzling the Eyes

Huike Wen

LEXINGTON BOOKS
Lanham • Boulder • New York • Toronto • Plymouth, UK

Published by Lexington Books
A wholly owned subsidiary of Rowman & Littlefield
4501 Forbes Boulevard, Suite 200, Lanham, Maryland 20706
www.rowman.com

10 Thornbury Road, Plymouth PL6 7PP, United Kingdom

British Library Cataloguing in Publication Information Available

Library of Congress Cataloging-in-Publication Data

Wen, Huike, 1977-
Television and the modernization ideal in 1980s China : dazzling the eyes / Huike Wen.
pages cm
Includes bibliographical references and index.
ISBN 978-0-7391-7886-7 (cloth : alkaline paper) -- ISBN 978-0-7391-7887-4 (electronic) 1. Television--Social aspects--China--History--20th century. 2. Television broadcasting--China--History--20th century. 3. Visual communication--Social aspects--China--History--20th century. 4. Mass media--Social aspects--China--History--20th century. 5. Social change--China--History--20th century. 6. China--Social conditions--1976-2000. 7. China--Intellectual life--1976- 8. China--Civilization--1976-2002. I. Title.
PN1992.3.C6W45 2014
791.45095109'048--dc23
2013031525

™ The paper used in this publication meets the minimum requirements of American National Standard for Information Sciences—Permanence of Paper for Printed Library Materials, ANSI/NISO Z39.48-1992.

Printed in the United States of America

To My Parents and Brother

谨以此书献给我的父母和兄长

CONTENTS

INTRODUCTION

It was not until 1958, thirty-one years after the first television set was created, that China produced its first TV set and set up its first television broadcasting facility, Beijing TV Station. Because the Chinese political and economic system defines television, like all other media, as a tool of education and propaganda, TV's development was discouraged in the 1960s and 1970s. But in the 1980s, TV-set ownership grew rapidly, increasing from 4,850,000 in 1979 to 2,7,610,000 in 1982 (Guo, 1997). Although the number of people who owned a TV set was small in comparison to China's overall population, it was obvious that TV sets were becoming very important pieces of equipment that individual families wanted to own. During this time the number of Chinese TV stations also was increasing, and Chinese Central Broadcasting University was established in 1979. At the same time, magazines and newspapers relevant to TV were created in the early 1980s and the quantity and variety of TV programs constantly expanded (ibid.).

Most studies of Chinese television focus on the 1980s and later, when China started opening up to the world. Chinese TV in the 1980s is unavoidably connected with China's political and economic status in the world; therefore, most scholars outside China, as well as most overseas-educated Chinese scholars, concentrate on examining how political and economic power and cultural factors have interacted with Chinese policy in the television industry and administration (Lee, 2003; Chang, 1989; Chang, 2002; Hong, 1998; Curtin, 2007; Donald, Keane, and Yin, 2002; Zhao, 1998). On the other hand, because of the practical orientation of Chinese intellectuals, Chinese domestic scholars are often most interested in how to regulate TV and help the administrative institutions; they have almost always focused on the three major functions the Chinese government wanted television to serve: propaganda, education, and cultural enrichment (Huang and Yu, 2003). Many scholars of the history of Chinese television have followed the top-down format based on the political situation in China (Zeng, 2002; Yang, 2003; Gao and Wu, 2002; Hu, 2000; Yin, 2002; He, 2005). International and domestic scholars' agendas are often opposed. For example, international scholars are more critical whereas Chinese scholars are more interested in how to help develop Chinese TV and create an internationalized Chinese TV industry. Despite the different approaches, however, the larger topic is the same: how economic reform has influenced the Chinese television industry and given it a place in the international media industry. As David Zweig (2002) argues in his analysis of China's internationalizing process, "China's story speaks to key issues of interest to generalists in international political economy" (p. 6).

The preponderance of academic studies on Chinese TV shows that both domestic and international TV scholars have realized the importance of Chinese media in a globalized world. Although studying Chinese television might be less informative than studying television in other countries because of the prevalent censorship in China, the biggest audience in the world nevertheless makes the topic important for international scholars. However, the general study of TV and culture in Chinese society is still based on the assumption that television both influences and reflects people's lives (Zeng, 2002; Yang, 2003; Gao and Wu, 2002; Hu, 2000; Yin, 2002; He, 2005). Thus, to better understand television's role in modern Chinese society, I want to approach the relationship between Chinese television and globalization from a perspective that so far has not been considered sufficiently. My book investigates the Chinese government's official policy during the 1980s on television and modernization, more specifically on the promotion of TV sets, but also focuses on how the Chinese media represented television through coded images. The study is about the representation of intellectual reception of TV in 1980s Chinese media. Scholars have already generated statistical evidence about the Chinese television industry, examined governmental policies that were stated clearly in policy documents, and disclosed the effect of the international media industry and market, but these studies are mostly based on official political and economic records. My book aims to recover the elided records of Chinese television history, the images themselves. By examining the images of and on TV in more relaxed (at least superficially) cultural forms, such as "popular" magazines, advertisements, TV shows, newspapers, and films, I seek to explain "what various media institutions assumed about the public's concerns and desires" (Spigel, 1992, p. 8) and what the government as well as media institutions wanted the public to be concerned about and desire regarding television in 1980s China.

Learning Methodology from Lynn Spigel

In *Make Room for TV*, Lynn Spigel (1992) proposes a very illuminating way to think about television's relationship to culture. Her research engages two major issues about which scholars have argued for decades and that are still very important in media studies. One regards the best way to examine the relationship between different media genres. Spigel's study of television and the American family ideal is based on her examination of four popular women's magazines. From the magazines, Spigel extracted advertisements about television, to which other scholars of television had not paid enough attention. Spigel shows how the advertisements reflect the dominant post-World War II ideologies of family, work, and gender roles. As a different media genre, magazine advertisements provide a way to look at television's role in contemporary society. Spigel's study is a concrete example of a communication studies technique by which scholars consider one genre with the information provided by another.

Two of Spigel's contentions inspire my study. The first is that academic studies, especially historical studies, are always based on information provided by other media with a broader view, such as books, journals, biographical records, and many other reference resources. The second is that two of the main genres of mass media, television and magazines, that usually are treated as mere entertainment should be considered as fit subjects for academic study. *Make Room for TV* contributes to our understanding of the economic, political, and cultural elements that are connected with and reflected in the media's representation of the relationship between ideal family life and television. Spigel still focuses on the "top" level of the examination, rather than on reception research, but her study provides a way to look at the intersection between audiences and television. She shows how the dominant ideology spreads, through images, its idea of the ideal family. Spigel's research is not a pure study from a political economy perspective, which mainly focuses on the examination of governmental policies and the structure of the media industry. Nor is it a typical study of TV's role in people's lives. It is a study in between, which is partial, as Spigel admits, and it bridges studies that focus exclusively on either industry and government or on audiences. The advertisements in the magazines she considered were, of course, not produced without a careful examination of what the people wanted in the post-war America, but at the same time did not ignore the family ideal articulated by the wider culture.

The other major issue that Spigel's study contributes to media studies is her consideration of television as both a technology and a carrier of thoughts. She begins her study by examining TV sets as a type of equipment a family must own in post-war America. Spigel provides a way to look at how television is situated within a broad cultural environment. *Make Room for TV* examines the wish for an ideal family propagated by the dominant political and economic powers. We can question the degree to which Spigel's selection of advertisements is a representative sample; it is possible that her study has a selection bias and some advertisements in women's magazines may have presented views that conflict with the "facts" in her book. However, it is impossible for any scholar to do an exhaustive historical study at this or any point. Spigel is very conscious of being not impartial, and being impartial is the risk that scholars must take.

William Uricchio's (2004) study should be valued as another important examination of television, space, and cultural ideals in the intersection between the media industry and audiences (pp. 163-182). The studies contrast in some ways. Spigel's study concluded with the 1980s and Uricchio's concluded with the

1990s; Uricchio's is an empirical examination of the physical space of TV, while Spigel's is a study of the media representation of the relationship between TV and the family ideal. But their studies share a focus on textual traces of the negotiation between industry understanding and audience desires.

The studies of Spigel, Uricchio, and other scholars have provided the models by which I will examine television and culture in China. As a nation that is tightly controlled by the government, China is a good example of how a dominant central power might both understand and educate its citizens through television. In Michael Curtin's (2007) outstanding exploration of Chinese media, he attributes the fast dissemination of television in 1980s China to President Deng Xiaoping's promotion of the "Four Modernizations" (p. 192). Television, as one of the most important representations of modernization, was certainly connected with Chinese governmental policy (Zeng, 2002; Yang, 2003; Gao and Wu, 2002; Hu, 2000; Yin, 2002; He, 2005; Latham, 2007; Hong, 1998; Curtin, 2007). The 1980s constitute the early period of Chinese economic reform. In the 1980s, political and economic strategies were experiencing a dramatic transformation after the Cultural Revolution in China; political ideas and economic orientations were entangled and enforced, as well as in conflict with each other (Chu, 1978; Hong, 1998; Latham, 2007; Curtin, 2007). Although scholars such as Latham (2007) assert the importance of TV-set ownership to Chinese modernity in the 1980s (pp. 50-53), many cultural studies scholars fail to consider how technology became necessary in people's lives after the Economic Reform, despite the well-known fact that before the Economic Reform, most technology, including TV, was inaccessible to most Chinese people.

A study of Chinese television from the same perspectives used by Spigel is rewarding for two reasons: first, the government, the unquestionable decision-maker in China, creates and supervises the tone of media based on its economic conditioning; second, Chinese people are accustomed to being both educated and alarmed by the ideology of the dominant power. If this tractability of Chinese people was beginning to lessen in the 1980s, the change was still limited to a small segment of the population and in fact traces of it still exist due to the political structure and the common practice of governmental control. Thus, an examination of how the media promoted acceptance of TV in Chinese society leads to a better understanding of how imported Western technology was embedded in people's lives. It is also important as a window into the interaction between "old" local culture and the newness of the globalized world. This study is also helpful in gaining a deeper understanding about an alternative model of modernization. For the above purposes, this book explores the following questions from the perspective initiated by Spigel: What strategies do the media employ to persuade people to buy and watch TV? How is TV connected with modernization and individual life in these promotions? And how are people and space connected in the articulation of TV ownership in the 1980s?

Methodology

This book is a historical-critical study supplemented by a few oral-history interviews. As a relatively new media genre, television is a fresh memory to many Chinese people, given its debut in common people's lives in the late 1970s and early 1980s. To explore the media's imagination and construction of the relationship between TV sets and everyday life, it is important to learn people's thoughts about why it was necessary for them to buy their first TV set. The interviews were designed to examine whether people had a common purpose when they bought their first TV set and whether their purpose echoed the media's messages concerning TV-set ownership. I received surprisingly diverse answers from the interviews. I interviewed twenty-eight people of both genders and from various backgrounds, asking them two questions: (1) When did you or your family first acquire a TV set? and (2) Why did you or your family want to buy a TV set at that time? The responses I collected illuminate my critical examination of the images published in the periodicals I explore.

I focused on the following periodicals: *Life out of 8 Hours*, *Modern Family*, *Chinese Advertisements*, *Chinese Pictorial*, *Shanghai Pictorial*, *Popular Cinema*, and *Popular TV*. With the exception of *Chinese Pictorial* and *Popular Cinema*, all debuted in the 1980s. *Life out of 8 hours* was the first Chinese magazine created for leisure in China and, at its peak in the 1980s, circulation was more than 1.4 million per month. Launched in 1985, *Modern Family* was one of the most well-known magazines to target female readers. *Modern Family* was originally sponsored by Shanghai Women Association, which made it an important resource for studies on the relationship between modernization, family and gender. *Chinese Advertisements*, created in 1981, was China's first advertising magazine and therefore an important source for the study of TV advertisements. *Chinese Pictorial* (founded in 1950) was published in Beijing, where politics was a dominant concern, *and Shanghai Pictorial* (founded in 1982) was published in Shanghai, where economic reform was a higher concern. They were both very well known. *Popular Cinema* and *Popular TV* are similarly familiar to most Chinese readers. *Popular Cinema*, created in the 1950s, was one of the most popular leisure magazines in Chinese households in the 1980s and it played the preeminent role in providing commentary on films and TV programs in the 1980s. *Popular TV* was born by TV in 1980 and has served TV industry since then.

To compare the representation of television in the above leisure magazines and those in the newspapers and magazines directly controlled by the Chinese Communist Party (CCP) and clearly labeled as the tongue and throat of the state, I have selected *People's Daily* and *Fortnightly Chat*, two very important periodical presses of the CCP.

Literature Review

Although my study did not examine images exclusively, the bulk of materials I considered ultimately relate to an understanding of images. Therefore, it was important for me to have a methodology for reading images. Images can be very complex and invite competing interpretations because of viewers' different perspectives. "Images generate meanings" (Sturken and Cartwright, 2001, p. 45). "Most if not all images have a meaning that is preferred by their producers . . ." However, "we usually have no way of knowing for certain what a producer intended his or her image to mean" (p. 46). This ambiguity makes images different from spoken and written language, but challenges the extent to which any examination of images can achieve certainty. Therefore, I will position my book in relation to the literature of technology studies and studies of gender and mass media and of globalization and culture. Technology studies not only helped me investigate TV as a technological product but also provided a philosophical framework in which to look at media in their cultural context. The literature on gender and mass media enabled me to examine images critically to understand their significant meanings. Finally, because this book is a historical study focused on the 1980s when globalization haunted national cultures through, especially, the influence of mass media, I will explore the articulation of TV in Chinese media interacting with those of other nations.

Technology Studies

Numerous approaches from a body of diligent scholars have made technology studies a dazzling and complicated field. Nonetheless, it is possible to map the major divisions if we draw a line between the study of technology in general and the study of media technology specifically. The two areas define "technology" from humanist and sociological perspective: the former draws attention to what the general understanding of "technology" is, and the latter extends the common understanding of "media."

Lewis Mumford is one of the main scholars to take a macroscopic approach to studying technology, putting it into a historical and cultural context. He investigates technology and society through an integrated consideration of almost every element in "the social milieu—monasticism, capitalism, science, play, luxury, war—and the more specific achievements of the inventor, the industrialist, and the engineer," even the psychological and practical origins and esthetic and ethical results of machines (Mumford, 1934, pp. xi-xii). Mumford's theory of the relationship between technology and society is much clearer than those of his peers due to his choice of words in defining "technology." He uses the word "technics" in contrast to the word "machines" throughout *Technics and Civilization.* According to Mumford, technics, which are always part of the social milieu, arise from and interact with the invention and usage of machines, as well as being a result of the usage. The technics defined by Mumford can be understood as a kind of ideology, in the cultural studies sense of "ordinary ideas" (Carey, 1989, p. 210) but not the political understanding in Marx's definition. In Mumford's words, "the whole mode of life" has adapted to the pace and the capacities of the machine (1934, p. 4). For Mumford's contemporaries, the most provocative point in his study was that "ideas" have driven the accelerated development of Western society, specifically in Western Europe. Mumford synthesizes material and ideal factors in Marx. While it has long been believed that the process of social change is shaped by the fast and continuous invention of machines, it was not the steam engine, as Marx believed, that has been the basic driver of modern society. Instead, what drives society are the more basic but invisible "ideas" born in the process of living with machines. People's attempts at better, more efficient lives started long before the invention of the steam engine, for instance, in the mining industry. It was not the steam engine and its machine power that pushed the development of society; rather, the "monopoly and concentration" reinforced by the characteristics of the steam engine caused the appearance of "concentrations of buildings and peoples" in the new areas in the industrializing process (Mumford, 1934, pp. 161-164). Mumford emphasizes the technics and ideas embedded in the steam machine, which is different from Marx's concentration on the human relationship between people who own the steam machine and the people who work for the owners.

Stated differently, Mumford believes there are characteristics inherent in machines, but Marx pays more attention to the people who own the machines and have the power to control laborers through this ownership. For Marx, machines cannot change anything if their owners do not have the desire to control others and explore potential capital. The civilization in Mumford's description is harder to change than the one in Marx's. It is because Marx's society and civilization can be changed when the machines belong to everyone and the control disappears. However, Mumford's society and civilization will still exist unless human beings destroy the machines and throw away the "ideology" that machines have planted in human minds. In Mumford's view, the technics and ideas are ubiquitous and remain even after the machines that caused their existence have been replaced by or transformed into new machines. Mumford hints that the invisible ideas that machines carry with them are more powerful than the visible and tangible productive or constructive power brought by machines. Mumford's ideas touch almost every field in social science, but his early discussion gives no final answer to the classic question, How can the capitalist system let the fluidity of these ideas and technics exist without using or disturbing them?

Mumford tried to answer this question in a later work, *The Myth of the Machine*, demonstrating that technology could have made a healthier society if modern technology, which he calls megatechnics, was not steered by elements outside of technology production. He points out that "without constant enticement by advertising . . . production would slow down and level off to normal replacement demand. Otherwise many products could reach a plateau of efficient design which would call for only minimal changes from year to year." In other words, the problems existing in modern society demonstrated by megatechnics are not caused by modern technology; instead, it is *human agency* that leads to the unnecessary amounts of technology production. It is not hard to tell that Mumford was both an optimistic idealist of technology and a serious critic of the human agency that interrupts the development of technology.

Traces of Mumford's argument are found in the work of many other technology studies scholars. *The Control Revolution*, James Beniger's (1986) investigation of information technology, actually shares the same thesis: the phenomena caused by technology in contemporary society have been in place since long ago. Beniger advances Max Weber's philosophical thesis that bureaucracy is the most important control technology, but his scope and methodology are reminiscent of Mumford because Mumford also owes much to Weber. One of the concepts Beniger identifies as important to technology is the bureaucracy included within Mumford's concept of technics. However, the role of bureaucracy is exactly the aspect of technology that Mumford as well as many of the scholars following him did not sufficiently emphasize. In *The Control Revolution*, Beniger wrote, in a very accessible style, a history of information technology that is full of disruptions and crises. Beniger believes that the desire to control information was originally caused by industrial production, and that the bureaucracy had to find a way to organize over-produced products. However, once bureaucratic control started, the consequences were not directly driven by production; instead, production became circular, driven by controls of the bureaucracy. In Beniger's opinion, information technology and bureaucracy are both technologies that the capitalist system uses to resolve its crisis of control.

Actually, the causes of the control and the crisis in Beniger's argument are already accounted for in Mumford's technics. The strategies Beniger identifies as being used to attain control and resolve control crises only made adjustments to the technics. For example, Beniger argues that family relationships were not dominant in business before the invention of modern technology and communication, due to the separation of production and retail sales. As a result, people's relationships were rationalized. Based on Beniger's argument, we can tell that the control revolution cannot resolve the crisis of control in society unless technological innovation and commercial production as well as bureaucratic control are stopped, which, as we all know will not happen in the near future. To resolve capitalism's crisis of control, the system can only develop new "technologies," be they different types of bureaucracy or new information technology, and continue dealing with new crises as a result of resolution of the "older" crises.

If Beniger's argument suggests an ideal way for us to face the crisis of control, Borgmann's (1984) argument concerning "technology and the character of life" reminds us how impossible and unrealistic these methods would be. Although Borgmann never clearly references Mumford, Borgmann's work reflects Mumford's early study of the relationship between technology and living conditions. Borgmann echoes Mumford's opinion on the relationship between technology and the ideas of human beings and argues that those ideas actually come from human beings and are reinforced and embedded in human beings' minds after technological innovations fulfill people's most basic needs. Technology promises to help people escape from the "toil," and to a different degree fulfill it; however, at the same time, it makes people lazier about both physical work and mental work. People are going further into an effortless and frivolous life. And people seem to be continuing to do so, no matter how much we have realized that this type of dependence on technology causes problems. Borgmann, as a philosopher, brutally but honestly points out that human beings do not want to be victims of the crises caused by technology, but they cannot escape from it, because technology is already a permanent presence. For Borgmann, technology is a philosophy that is embedded into society and human beings' minds without anyone teaching it. Technology is a natural part of the unnatural and technologized us.

The general tone of many scholars is one of helplessness in facing the fact that technology is embedded in our lives. Marshall McLuhan and his followers switch gears. McLuhan wants to rouse human beings from their technologically caused hypnosis. McLuhan is famous for his bold statement that the "medium is the message" (McLuhan and Frank, 1995), which, combined with his "frivolous" (Peters, 2005) style, resulted in widespread misunderstanding of his work in the 1980s. It is an irony that McLuhan, who came from a literary background and explained media with quotations from poems and prose, was crowned the representative figure of technological determinism. Even though McLuhan clearly states that the medium is the message, he did not mean to argue that the message the audience gets from a medium is completely decided by the medium. Instead, McLuhan's point is better understood in light of Borgmann's analysis of the relationship between people and technology: technology's existence in our lives has become invisible because life with it has become habitual.

Paradoxically, McLuhan's emphasis on media technology's influence on the audience's acceptance of the message reflects exactly this habituation to technology. Watching television becomes a part of modern life, so that whatever meaning any particular program carries often does not really matter to the audience. It is the watching experience that matters. McLuhan, in fact, pays much attention to the message that a medium carries; according to him, programs repeatedly contain the same mundane content with little variation. In other words, an audience that has become accustomed to living with a certain medium is already familiar with the ways in which that medium communicates, and the general message the medium wants to transmit; thus, what is presented in the media is actually nothing different than what they already know.

But each genre of media, such as TV, films, and radio, has its own format for sending messages, and audiences notice these differences when they switch from one to another. McLuhan in part wants to draw people's attention to the medium itself instead of to the message presented through the medium. Kittler was also criticized because of his daring statement that "media determine our situation" (1999). Despite Kittler's interest in Foucault, Lacan, and many other thinkers in his analysis on the relationship between people and gramophone, film, and typewriter, some scholars could not get past the first sentence of his book and regarded him as a technological determinist. Actually, the fact that some scholars ignore McLuhan's and Kittler's intellectual considerations beyond technological determinism reflects, as John Peters points out, that technological determinism is an abusive term in the field of technology studies. But this so-called technological determinism cannot be found in the work of any of those scholars who are supposed to be proponents of it. Adherents of McLuhan and Kittler both enrich and create new branches in media studies. We can divide media studies before the 1980s into two tendencies: as John Peters proposes in his analysis of Innis and McLuhan, "one focused on social and political organization, the other on psychic and sensory integration" (Peters, 2005). The former concerns are more traditional, although Innis's approach was not traditional at all. However, the latter, which McLuhan is made to stand for, is somewhat distant from the usual concerns of social science and philosophy. What McLuhan, Innis, and also many Canadian communication scholars contribute to communication studies is that they direct attention to a detailed examination of specific media. They not only argue that media are part of the society but also treat media technology as the window through which to look at society.

More recent scholars have the legacy of McLuhan and Innis. In Walter Ong's (2002) media study, *Orality and Literacy*, we can see both Innis's scope and McLuhan's and Kittler's enthusiasm for the psychic and sensory integration of media and human bodies. Ong has managed to overcome the privileging of culture in his examination of "the differences in 'mentality' between oral and writing cultures" (2002, p. 3). Innis was the first to investigate media with this broad vision; he considered media to be anything that can transfer information, not just the concrete modern media technologies.

In Schivelbusch's (1986) *The Railway Journey*, as well, we can also see the considerations of space, time, and technology that Innis raised. Whether the scholars who wrote about media technology after McLuhan were directly inspired by him or just responding in similar ways to the dominant presence of technology, much scholarship has been done in this vein. Much of this has carried McLuhan's general ideas into domains by concentrating on the telegraph, the television, the radio, and sound technologies. Perhaps this is like what McLuhan explains, "one medium's content is always other media" (McLuhan and Frank, 1995, p. 151), or Innis's (1951) generalization that every medium is either space or time-biased; therefore, scholars spend much time comparing the different media. Comparisons among different media genres bring us to a deeper understanding of media technology. Ong's comparison between orality and literary, Bruce Owen's (1999) analysis of television and Internet, and Jeffery Sconce's (2000) tracing of television to telegraphy are all good examples. And comparisons of different media from both historical and contemporary perspectives have included provocative theses, such as Carolyn Marvin's:

> The history of media is never more or less than the history of their uses, which always lead us away from them to the social practices and conflicts they illuminate.
>
> New media, broadly understood to include the use of new communications technology for old or new purposes, new ways of using old technologies, and, in principle, all other possibilities for the exchange of social meaning, are always introduced into a pattern of tension created by the coexistence of old and new, which is far richer than any single medium that becomes a focus of interest because it is novel. (1988, p. 8)

Most scholars have contended that the history of media must be learned from the analysis of the history of their uses, as Mumford, Beniger and Marvin all argue; therefore, many recent scholars are especially careful about how the specific medium of their study has been used. Some of the most recent studies on media, such as acoustic technology studies, including Wurtzler's (2007) "electric sounds" and Emily Thompson's (2002) studies of "soundscape," consider how technology has been used and redesigned according to economic benefit and performance quality. Thompson's theory of how soundscapes have historically interplayed with modern architecture is a fascinating example of a study of how a medium has been designed. As a medium that incorporates facets of all the media technology that preceded it, television has become an important focus of technology studies. Television bridges older media, such as radio and film, and newer media, such as the Internet. Because television has become a popular appliance in most households, it is an important subject in studies of technology. TV has become a key focus in scholars' explorations of the character of life in a technological society, social and cultural constructions of media, the relationship between media and space, the transformations of media in an internationalized world, and a real case of "a medium in transition." People in early twenty-first century are living with TV, as well as somehow walking away from it. Represented by Raymond Williams' studies, the study of TV has been tightly connected with critical cultural studies. As he claimed, TV is a "technology and cultural form" (Willams, 2005). Williams' background in literature orients him toward textual and literary analysis, but he still pays much attention to the "form" shown on TV, not just the absolute "content." He treats TV as a technological and cultural form that is produced and controlled by the TV industry. Fiske and Hartley's classical work, *Reading Television,* shares similar views with Willams. They also direct attention to the TV industry and the larger cultural environment of which TV is part. They consider both the socio-political issues involved in producing TV programs and the difference between TV and other media genres. At the same time, they also explore the interactive relationship between TV and its audiences.

Williams's and Fiske and Hartley's studies inspired a generation of TV scholars, and the people who made TV studies a popular field work from a feminist perspective. Brunsdon (2003) argues that feminism went from being a political movement to a theoretical orientation when feminist scholars began studying soap operas in the 1970s. Through these and other studies, feminists have contributed to the academic community while remaining social activists. Feminist scholars' roles in TV studies are clearer when we see a long list of TV studies that include works by many female scholars, even though not all indentify as feminists, such as Brunsdon, Spigel, Hilmes, Whyte, Feuer, and D'Acci. Numerous TV programs have provided material with which female scholars have exposed ideological secrets of patriarchal society. At the same time, the argument of these scholars concerns the content of TV programs, which can be seen in the evolution of the gender roles affected by the feminists over the past few decades.

Many scholars of television echo Winner's (1993) call for social responsibility in technology studies. Winner criticizes "social constructivism" scholars for ignoring the ethnical "meaning" of their studies. Callon's (1987) article "Society in the Making" exemplifies those criticized by Winner. Winner argues that social constructivists are only interested in the origins of particular technologies and do not care about the technology's potential social consequences. They turn technology studies into pure studies of "technology," although they studied the intellectual activities that happened in the process of inventing a technology.

Most studies of technology share an interest in social and historical context; it is the basic if not the most important consideration in any topic in technology studies. This concern is displayed by recent studies of the Internet. Scholarship by Terranova (2004) and Jenkins (2006) differ greatly in the claims they make about the Internet, but they both maintain a sense of scholarly responsibility. Jenkins has been criticized for his optimistic view about people's agency in technological culture and for his casual writing style, but his political position is very clear—he wants to celebrate the power the new technology has provided to people. Terranova, on the contrary, admonishes us to think about whether "free labor" is really "free."

Studying media with the consideration of the changing social and cultural context is the direction that I take in my study of Chinese TV in the 1980s. In this book, I use Mumford's notion of technology and social milieu and think critically about the relationships among TV, audiences, media industry, states, and different genres of media discussed by this brilliant group of scholars that I mention above. I also borrow methodology from feminist scholars of television, among which Lynn Spigel's (1992) view is the most important to my study.

Issues Related to Globalization: Glocalization, Nationalism and Consumerism

Globalization is a much-studied topic and one to which scholars take different approaches, including historical, geographical, and cultural perspectives. Tomlinson's (1991) early examination of globalization focuses on cultural imperialism, a phenomenon which was later questioned by Appádurai (1996) and other postcolonial scholars who argue that influence between cultures is multidirectional. In Marwan Kraidy's (2005) case study of hybridity, he argues that hybridity can help us understand the influences among different cultures in the process of globalization but he does not ignore the uneven power dynamics described in Tomlinson's theory of cultural imperialism. Nancy Morris (2002) claims that globalization is not something new: no pure cultures exist; all cultures must learn from each other to survive (p. 278). At the same time, no matter how obvious the change to any culture, each culture has its own essence.

In spite of the different approaches on globalization and culture, it is commonly accepted that the recent globalization was initiated as a way of extending economic power in the world. Television, a technology invented in the West, and a carrier and distributor of cultural products embedded with various ideological elements, is an important target for scholars. Examinations of television and culture in China or in other specific places cannot ignore the fact that we are living in a globalized world. Many studies examine how Chinese media have been influenced by governmental economic and political policies made in response to both domestic and international situations. These studies of globalization form the background of my historical examination of Chinese TV and of the idealization of modern life by the Chinese government since the 1980s.

Studies of globalization help us examine the unique process through which TV was embedded in Chinese people's lives according to Chinese space, ideology, and general social environment. The theories of globalization create the foundation of my examination of TV in Chinese cultural context and encourage me to ask questions, for example, "How did the domestic producers advertise their TV products?" At the same time, the answers to this type of question are also connected with the localization of TV as a byproduct in the globalization trend. In their studies of the Internet in China, Maynard and Tian (2004) confirmed that glocalization is the outcome of the interaction between globalization and localization. Globalization was also in process at the time when TV was first introduced in China. Therefore, in order to gain a clear picture of how globalization was impacting Chinese local culture, an inquiry into the history of TV in China is very important.

Another very important issue related to TV in China is the relationship between nationalism and globalization. As is well known, the indoctrination of nationalism is always the center of education in the People's Republic of China (PRC). Nationalism was the main element in the PRC's propaganda in the Economic Reform too (Rosen, 2003; Chu, 1978; Chang, 1989; Hong, 1998). Because television was an important component of the governmental policy of "Four Modernizations," we should ask how TV fit within the complicated relationship between "traditional" nationalism, modernization, and globalization. Chinese cinema has been highly influenced by nationalism, as pointed out by Yingjin Zhang (2004) in his examination of Chinese cinema and culture in the early twentieth century. This book wants to examine if nationalism paradoxically aided the acceptance of TV. If the introduction of TV in China is as an example of glocalization, then what are the particulars of this case?

Japanese government at one point convinced people to increase their consumption to help the national economy (Iwabuchi, 2002; Starrs, 2001). This is one example of how nationalism is very much related to consumerism in modern society. The relationship between nationalism and consumerism is getting much more complicated in the gradually more complex situations created by globalization. TV has been influenced by this type of complexity in China. Through exploring TV and its installation in Chinese people's modern lives, we will be able to get a better understanding of this complexity. I want to explore if the promotion of TV sets brought new ideas about consumption to China, and if this change was related to the dominant idea that was looking forward to modernization at the time.

Gender, TV and Modernization

Many scholars have contributed to our understanding of the relationship between media and gender. Some have argued that gender roles in real life have been affected by gender representation in the media (D'Acci, 1994; Giles, 2007; Ussher, 1997; Bordo, 2004). Scholars have celebrated the media revolution of recent years in showing unconventional roles, such as gay and lesbian characters, powerful women characters, and people in nontraditional relationships on television, and in movies, and have also critiqued, as perverted, representations of certain gender displays.

Scholars focus mainly on textual analysis of media portrayals of gender. Gender studies scholars have offered critical interpretations that can be applied to my study of television in China. Lynn Spigel's *Make Room for TV* examines how gender roles were presented in advertisements for TV sets. Spigel's studies portray television as a medium that reinforces, as well as changes, traditional gender roles within the home as well as relationships among family members. Spigel builds her argument by examining how TV was portrayed in advertisements and historical records such as portrayals of people watching TV in TV dramas and cinema. She considers TV content to reflect the dominant opinion on the relationship between TV and family members' daily lives. In this sense, Spigel's study is a study of TV through the contents of TV.

Spigel's special contribution is not only to analyze the representations of gender and family in particular TV dramas but also to examine the media's portrayal of TV as important to the changes in gender roles and the relationships between family members. Through her research, we can get a closer look at what people in the United States thought during the 1950s about gender differences and the problems existing within the family.

Spigel's study discloses a fact ignored by many scholars, that the dominant powers with direct access to the content of commercials and TV programming assumed that television had changed, and would continue to change, people's lives. This assumption of the hegemony was reflected in the opinions they embedded in the advertisements and the TV dramas. But at the same time, her study shows that the dominant powers' assumption was partly based on what they got from both their imagination and research. The dominant powers also embedded their concept of the idealized American family in the various messages related to TV, such as the role that TV plays in family life, and the contents of TV dramas.

The mass media are often criticized as representing gender problematically. Susan Bordo (2004) argues that media influence women's lives even when they attempt to promote an ideal of the "strong woman." When an ideology of patriarchal domination is present, even radical behaviors and thoughts of women can be appropriated by patriarchal society. Anorexia used to be understood as the tragic result of the impossible aesthetic criteria of consumer culture within patriarchal society. However, Bordo argues that underneath this seemingly yielding behavior is a rebellion against being controlled. Anorexic women actually want to be in control of their lives and to be strong. However, their good intentions lead them in the opposite direction because of their lack of a real revolutionary means. Bordo suggests that people look beneath the surface and determine the real ideology supporting societal gender roles. To simply assume that women are passive or subordinate to men and that men are active or dominant overstates the case. However, to simply believe that women are getting stronger and more powerful, and that men and women are equal at all things, is not accurate either. Gender issues are very complicated, and the representation of gender in the media is similarly complex.

Many scholars have performed detailed analyses of representations of gender on TV in China (Zhao and Zheng, 2006; Onishi, 2006; Li, 2001). However, these scholars fail to consider the issue from a historical perspective or within the extensive context of contemporary Chinese TV.

A Brief Review of the Definition of Modernity

How to define modernity is a hard question, but answering it is essential for a better understanding of the points raised in this book. From the sociological perspective of Anthony Giddens, "inherent in the idea of modernity is a contrast with tradition" (Giddens, 1990, p. 36). According to Giddens, modernity is characterized, mainly, by institutions, including capitalism, industrialism, surveillance, and military power. These institutions construct every modern nation-state and together create, regulate, and maintain all of modern society. They account for the economy, the environment, production, and science, and, therefore, modern culture in general. Although Giddens categorized the institutions into these four types, they are always entangled. According to Giddens, the biggest difference between tradition and modernity is that tradition has "the inertia of habit," but modernity always welcomes the "light of incoming knowledge" (p. 38). Modernity does not deny or completely exclude tradition. "[E]ven in the most modernized of modern societies, tradition continues to play a role." However, "tradition can be justified" in modern society, "but only in the light of knowledge which is not itself authenticated by tradition." Differing from modern society, "however, in pre-modern civilizations reflexivity is still largely limited to the reinterpretation and clarification of tradition, such that in the scales of time the side of the 'past' is much more heavily weighed down than that of the 'future'" (p. 38). According to Giddens, writing helps maintain tradition, because "literacy is the monopoly of the few," and "the routinization of daily life remains bound up with tradition in the old sense" (pp. 37-38). At this point, it is worth remembering that Chinese culture has always emphasized the superiority of literacy over other skills. This can explain why Chinese culture can stay in touch with tradition over its long history; intellectuals have always been very important in maintaining and inheriting cultural traditions.

Giddens and other scholars' arguments on modernity, on how the relatively traditional content in a culture interact with the newness, illuminate how I should look at the material that reflects how the newness has interacted with the essence of the older cultural content in 1980s China. Television, because of its characteristics as a technology and as a carrier of ideas invented in the West, serves as the nexus of my examination in this book.

Chapter Outline

The work of the aforementioned scholars informs my study of how television became a symbol of modern life in China in the 1980s and how television has interacted with and shaped this modernizing culture. I pay attention to the "technics" or "social milieu" when I analyze the images from magazines, TV dramas, films, and other visual materials.

Chapter 1, "What Could You Do with a Television Set?: The Legitimization of Television," examines how the Chinese media introduced and represented television in the 1980s. I compare how Chinese institutions promoted TV sets with the methods of American marketers. I attempt to flesh out the complicated and often awkward relationship between the values of traditional Chinese culture and the values promoted by TV advertisements in the 1980s. The issues I cover include, but are not limited to, filial piety, the education of children, the role of TV in the relationship between hosts and guests, and the complexities of materialization and intellectualization as reflected in advertisements' representations of TV in China at the time.

Chapter 2, "Jump Starting a Learning Process: The Ownership of TV Sets and Social Status," investigates how TV was presented by 1980s Chinese media as a symbol of modern life, wealth, and higher social status, and the idealized meanings of TV presented by the media about who owned a TV set in Chinese society. In this chapter, I investigate how other media genres, mainly print media, made TV disseminate a "new" cultural value—being modernized and becoming rich—that conflicted with economic and political reality.

In chapter 3, "Documenting Nature: TV and the Representation of Nature in 1980s China," I examine how television, a modern medium, has been connected by print media with nature from its inception. From advertisements that aim to promote TV sets, to the TV documentaries that occupy the biggest contingent of domestic-produced TV programs, the representation of the conflict between modernization and nature has been central. In this chapter, I analyze the well-known TV documentary *Talking about the Changjiang River*. By comparing this documentary with other TV documentaries that were produced during the same period, and analyzing the various aspects of the social context, I want to raise questions about the goals of these TV documentaries. This chapter also reviews traditional Chinese understanding of nature and how this has changed with the evolution of social and cultural ideology.

Chapter 4, "TV and the Other Modern Mass Media," concentrates on the relationship between TV and other media technology, such as film and print media. I explore how the ideology behind earlier media influenced people's opinions concerning the role TV should play in their lives. I also examine how governmental policy interacted with TV production and audience's responses, thereby making Chinese TV into a cultural form. I follow other scholars in tracing the transformation between media in public and private spaces. My intent is to explain what the cultural transformation was at that time in China, and to use the cultural elements in the brief cultural history of TV to help answer the question.

In chapter 5, "Providing More Choices?: Domestic and Imported TV Dramas and Gender Confusion," I focus on how the promotion of TV and the images of and narration about people in TV dramas and in TV-related media have represented gender in specific ways. I also examine the ways in which television interacted with other political, educational, and cultural issues to affect gender, and especially female gender identity in the 1980s.

My project as a whole supports my thesis that Chinese TV acted as nothing more than a "tongue" of the government when it aired domestically produced programs, and as a fantasy of an exotic culture or a modernized future for Chinese audiences when it narrated imported dramas from Taiwan, Hong Kong, Japan, Brazil, Britain, and the United States. I will analyze how the images, language, and stories in the latter present a pure virtual life and create an imagination of modern life for Chinese audiences.

In the epilogue, I briefly review the general situation of Chinese television since the 1990s. The review serves as a comparison of TV as a newcomer and a "natural" participant at different stages in the process of modernization in China. The epilogue also clarifies my studies on media technology and globalization, how other media promote this relatively new media, and how they learn from Western experiences with promoting new media technologies and visual products, and what local promotion strategies they have invented in this process.

CHAPTER 1
WHAT COULD YOU DO WITH A TELEVISION SET?: THE LEGITIMIZATION OF TELEVISION

> In its most original meaning of "transmitting" and "communicating," tradition does the work both of preserving and of transcending, as is found in the processes of scientific growth.
>
> *TRAN VAN DOAN, Chinese Cultural Traditions and Modernization (Wang, Yu, and Mclean, 1997)*

China began modernization hundreds of years ago. But large-scale Western-style modernization did not begin until the nineteenth century, when China started interacting with Western countries. This process can be traced back to the time of the First Opium War, in 1840 (He, 2002, pp. 4-5). Western technology and weapons used in this war scared many Chinese people. At the same time, the British army's victory and changes brought to China by Western culture signified the power of a new political, economic, and military system. Many upper-class Chinese intellectuals shared an eagerness to reform and strengthen China by adopting Western technology.

The modernization process in China has frequently struggled between the desire to protect the nation's political system, which is different from Western political systems, and the selective imitation of other aspects of Western culture, especially modern technology and science. For example, during the Qing Dynasty, the ruling class wanted to adopt Western technology and strengthen the country's national defenses but also wanted to maintain the central feudal dynasty's control. However, as has been seen throughout history, once a country's door is opened to outside influences, what enters is not as easily controlled as the dominant power might want. Technology is not just about machinery; technology encompasses philosophical and intellectual factors as well. Consequently, modernization, as many scholars have pointed out, occurs along many dimensions, including the technological, economic, social, political, and ethical dimensions, as Hirai Naofusa (1983) has argued regarding Japanese modernization. Value systems also need to be considered. Modernization and the understanding of tradition have been evolving since the dramatic change caused by the conflict between the old and the new. The modern state-nation relationship has been a focus of renewal and reform since the early twentieth century. Common people sometimes forget that socialist China itself is also a result of modernization, via the adoption of Marxist ideology and the Russian model and the rejection of the Republic of China (1911-1949). Thus, China's modernization in the 1980s did not begin in the 1980s, as many people think; rather "tradition" and "modernization" must be viewed through a longer historical lens from a broader historical angle.

Consequently, a complex question about China's modernization in the 1980s arises. Because it is impossible to talk about Chinese modernization without comparing it to Chinese tradition and Western modernization, which tradition should one focus on? Examination of media representations of television sets in the 1980s reveals the existence of central themes that have always been present in the Chinese culture's discourse concerning modernization, themes such as filial piety, education, and etiquette, as well as how to balance the pursuit of material satisfaction with moral improvement. A detailed examination of the continual evolution of these themes is beyond the scope of this book; however, I want to briefly examine the treatment of the four predominant themes of modernization in the context of the representation of TV in mass media in the period directly following Mao's socialist rule. The examination is based on a comparison between the explanation of the themes in Mao's Socialist China and the early post-Mao era—the 1980s.

This chapter focuses on how the media, using the four themes, portrayed the role of television, and how this portrait interacted with "traditional" values and created a contradictory discourse that was essential in the transformation of Chinese society in the 1980s.

In the Name of Filial Piety

The Cultural Revolution affected many traditional ways of thinking in Chinese culture, although the country remained unchanged in its fundamental beliefs. Among the most important beliefs, and one that is at the center of morality and social relationships, is "filial piety." Filial piety, called *Xiao* or 孝 in Chinese, has been mandated by many classical guidelines over the centuries (Ikels, 2004, p. 4). "In the classics and in popular thought, support, subordination (or obedience), and continuing the family line have all been touted as the essence of filial piety" (Ikels, 2004, p. 3). In the Chinese literary tradition, the foundational instruction that has shaped people's adherence to filial piety is *The Classic of Filial Piety*, in which Confucius said, "In serving his parents, a filial son reveres them in daily life; he makes them happy while he nourishes them; he takes anxious care of them in sickness; he shows great sorrow over their death; and he sacrifices to them with solemnity" (Chai and Chai, 1965, p. 331). Mencius (372-289 BC), the most important Confucian other than Confucius himself, further emphasized that the most unfilial behavior is "having no posterity." Charlotte Ikels has pointed out that the scriptural disciplines have been enriched with newer contents many times over the centuries (Ikels, 2004, p. 4). People are motivated to demonstrate filial piety by many factors, among which the most common is the belief that "the individual who has been trained well accepts the willingness to practice filial behavior as a key indicator of a mature, well-adjusted adult" (Ikels, 2004, p. 5). Although the scriptural tradition has changed, filial piety was always essential in pre-modern China to maintaining the stable patriarchal Chinese feudal society. Filial piety is the primary moral requirement of Chinese culture both historically and today. According to Kwang-Kuo Hwang (1999), filial piety is still essential in family relationships in Chinese culture. Chinese communist policy "eliminated the family as a production unit, made family property ownership and inheritance inconsequential, made social mobility dependent upon education and bureaucratic allocation to jobs, provided secure wages and a range of fringe benefits (including pensions) for most urbanites, and made cramped public housing available at nominal cost through bureaucratic allocation" (Whyte, 2004, p. 124). However, as pointed out by Whyte, it was not hard for grown children to provide filial support in China, even in the 1990s: because of the extreme shortage of housing in China, young adults had to live with their parents, who usually also received pensions from the government under the socialist system (ibid.).

Thus, communism/socialism in China changed the traditional way of performing "filial piety" but did not eliminate it. As Kwang-Kuo Hwang (1999, p. 179) notes, "although the content of Chinese familism has changed, cultural ideas about family are resistant to change." Dramatic changes in post-socialist 1980s China caused differences in values among people of different generations. Economic reform changed the "stasis" of social position individuals had in socialism into a fluid process of modernization. Younger people got the opportunity to change and improve their lives and increase their income, whereas older people were ready to retire or already quit the passionate pursuit of newness when they went into their older age. Therefore, it was not a big surprise to learn in the interviews that many Chinese people bought their first TV set in the 1980s, not for themselves but for their parents. For example, one interviewee, a professor who teaches at a private college in the United States and has become a permanent resident of the United States, said, "I had some extra money after I studied in the U.S. The dollar was very strong at that time. So after saving carefully, I bought a TV set for my parents when I visited them two years later. It was a 12-inch TV set, white and black. It was not fancy, but it was such a precious property for my parents, and all the people in the village which I grew up in got the chance to watch TV because of me." Buying a TV set was not the only way for children to show filial piety to their parents, but it was the best way, in the interviewee's opinion, for the parents to enjoy the honor of having filial, caring, and successful children.

Many interviewees told similar stories. They thought purchasing a TV set was a good way to bring some happiness to the lives of their elderly parents, especially those parents who had lost their spouses. One interviewee said, "The first TV set in our family was the one that my father and my uncles bought for my grandma, which made her very happy and many people went to her place watching with her. She was living by herself at that time. My father and uncles thought it was too lonely for her to live by herself. So they thought of buying her a TV." In this case, the television set was treated as consolation and company for the elders. The traditional way of performing "filial piety" through showing obedience to parents was modified in the modernized Asia (Sorensen and Kim, 2004; Whyte, 2004; Hwang, 1999; Sung, 1998). Buying a TV set to provide company to parents became a new filial piety practice in the Economic Reform China. Other Asian nations that value filial piety have had similar experiences. Sorensen and Kim (2004) pointed out that in Korea, the performance of filial piety has always been based on social status, and in modern times, economically and socially superior families have explored the new way to perform "filial piety."

> It is these families who consider purchased commodities an adequate means of performing filial piety—the affluent and educated—who are suspected of insincerity and shallowness in their performance of filial piety. (Sorensen and Kim, 2004, p. 180)

Sorensen and Kim's study illustrates the relationship between modernization and performance of filial piety; and Sorensen and Kim's study helps explain the new interpretation of this traditional value in 1980s China. Although the new ways of performing filial piety might be considered insincere and shallow from a traditional perspective, they nevertheless serve to carry on the value. The criticism of this new practice focuses on the changes that people have made to adapt this tradition to their busy lifestyles and their ability to take good care of their parents in a modern culture driven by consumption and production.

Likening the purchasing of TV sets for parents to the tradition of showing filial piety sets up a new reference point in the moral evaluation of people's behavior. Efforts in the periodical press to improve the moral standards of Chinese people often used images of TV sets and elders. For example, the cartoon in Figure 1.1 portrays young parents watching TV while the mother of one of them is washing clothes behind the huge sofa on which they are sitting. The old woman must work very hard and cannot even get a peek at the TV because the sofa completely blocks her view. Her grandson thinks she is exactly like the robot shown on the screen because both the woman and the robot work so hard. This cartoon is a satire of young people who did not take good care of their parents; it criticized the young couple for shedding the traditional virtue and suggested that in the future their son would treat them in the same way as they treated his grandmother. The denotation of the cartoon was obvious; what is perhaps not so obvious is that TV as the center piece of this criticism of the younger generations' failure of filial piety reflects the special role that TV was playing at that historical moment. It became a new criterion to look at when evaluating the moral standards of young people: were they watching TV while their parents were not, because the parents did not have TV sets or had to do housework?

Many portraits of the relationship between people and TV sets seem to answer the question above. In a comic published in *Life out of 8 Hours*, issue 4, 1983 (see Figure 1.2), a young couple goes to visit their parents and takes along a little gift, but when they leave, they have gained weight while their parents have become very skinny. More striking in this picture is that the young couple is shown carrying a TV set: They have taken the TV set from their parents' home. Whether the couple bought the TV set or it was paid for by the parents is not easy to tell from the picture, and the distinction does not seem important to the author either, because the most powerful element in this sharp criticism is that the couple "took away" the TV set from the parents' house. The shocked faces of the parents as they watch their children leaving the house with the TV set undoubtedly were intended to reveal the young couple's moral depravity. TV, as a prestigious product of technology, symbolizes the owner's financial situation, which I will come back to later in Chapter 2.

At the same time, TV is different from other types of acoustic entertainment equipment, which can be enjoyed by a bigger group of people while they continue to do their family work. The cost and the technological characteristics of TV sets made it unavoidable that while certain people had the privilege of watching TV, others did not. The cartoons in Chinese journals in the 1980s suggested that who could afford a TV set and who had the time to watch it became a window through which to view the moral uprightness of people. The textual construction in these cartoons is strong.

Ironically, older people's happiness could not be easily measured by whether they owned this expensive and rare machine, as the media frequently displayed. What could TV really bring into the lives of older people? Although the periodical press did not want to answer this question specifically, one can sense the conflict between TV's existence and the real circumstances *of these* parents' lives. Images published in Chinese magazines and newspapers in the 1980s conveyed the message that the presence of TV in older people's lives did not necessarily equal happiness. It seems that if a nice child, despite his or her own limited income and increasing financial responsibilities, gave a TV set to his or her parents, the parents should have been happy. However, the cartoons of the 1980s and early 1990s magazines and newspaper (see Figure 1.3 and Figure 1.4) often showed an old man sitting in front of a TV set and watching it alone. Although these comics were intended to be humorous, they also reflected a real aspect of elderly people's lives during China's modernization: TV was an important source of company, especially if the older people were single. Titled with "an indigenous remote," the cartoon showing an old man happily making use of a duster to switch TV channels from his chair is supposed to make the readers laugh (see Figure 1.3). But the repeated depictions of a man coexisting with a TV set and nothing else to some degree reflected the problem of the discrepancy between the symbolic meaning of owning a TV set in an old person's life and what "filial piety" was in the 1980s. What was more important for young Chinese people: keeping their parents company themselves or giving them a TV set to keep them company? This question was probably seldom thought about at the time, given that the elders shown watching TV in the picture always have happy expressions.

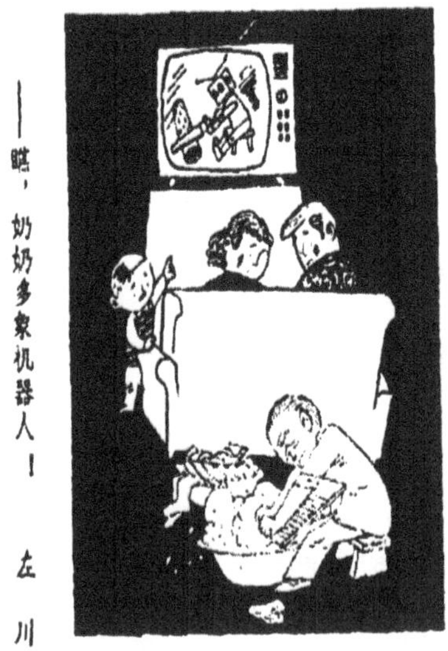

Figure 1.1: *Life out of 8 Hours*, 1983, no. 3.

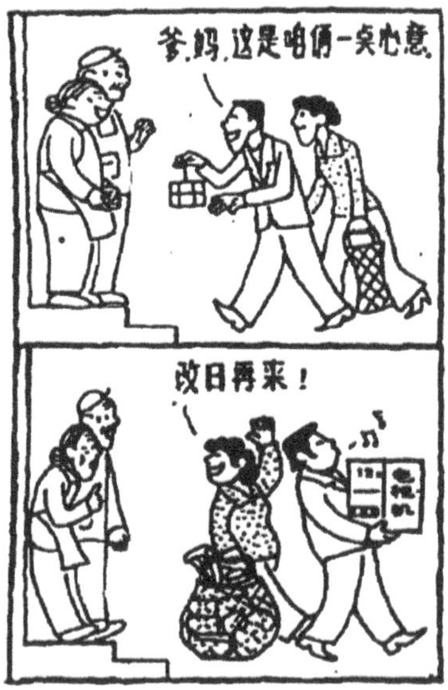

Figure 1.2: *Life out of 8 Hours*, 1983, no. 4.

In another cartoon (see Figure 1.4), the pleasant portrait of "how my grandpa tells what time it is" traces an old man's technological evolution, from telling time first by the sun, then by a clock, and finally by a TV set. The old man looks serene sitting in front of the TV set although there is only a test pattern on the screen. The still mood and the happy face shown in the picture indicate a change in the man's daily life. Waiting in front of the TV set before broadcasting has even begun for the day conveys the deeper meaning of the TV set in senior people's lives. As depicted in the cartoon in a comic tone, the old man and the clock on the TV represented the eagerness and happiness with which China was looking forward to a new modernization.

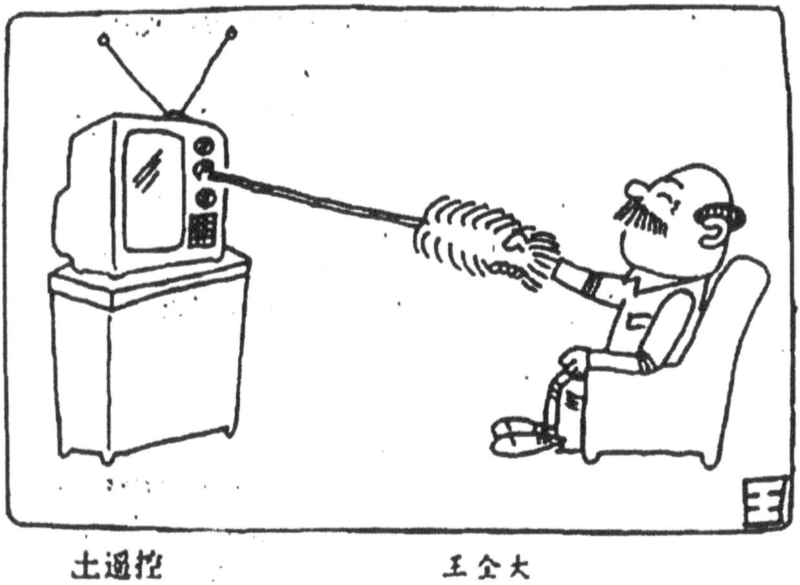

Figure 1.3: *Life out of 8 Hours*, 1990, no. 7.

The cartoon's creator did not realize the conflict in the scene shown in the picture: if the man told time by relying on TV signals, how did he know when to turn on the TV? Also, does the man's reliance on the TV set mean that he would leave the set on all the time, so he always knows what time it is? Or will his life follow the broadcasting schedule of the TV station? The TV set in this cartoon is both a substitute for other time technology and a replacement for things missing from the old person's life. In this cartoon, the TV set is portrayed as one installment in a long series of modern technologies: the calendar, the clock, and the TV set. The lineage depicted by the comics suggests that people's sense of time was becoming more and more regimented in modern life, but at the same time this invisible regulation was becoming even harder to notice in a society operated by modern "technics."

Unlike young people, whose time was still controlled by exterior entities, such as governmental institutions, state-owned companies, and other authorities, retired people often struggled to adjust to a solitary life without a time clock. The television set served as an artificial person and public in these elders' lives, as a replacement for children, coworkers, and other people the elders spent time with when they were younger. Retired people were happy when they were kept company by this "robot" which could fill in the space with sound and voices from all different backgrounds and display "living" people and their actions right in their own homes.

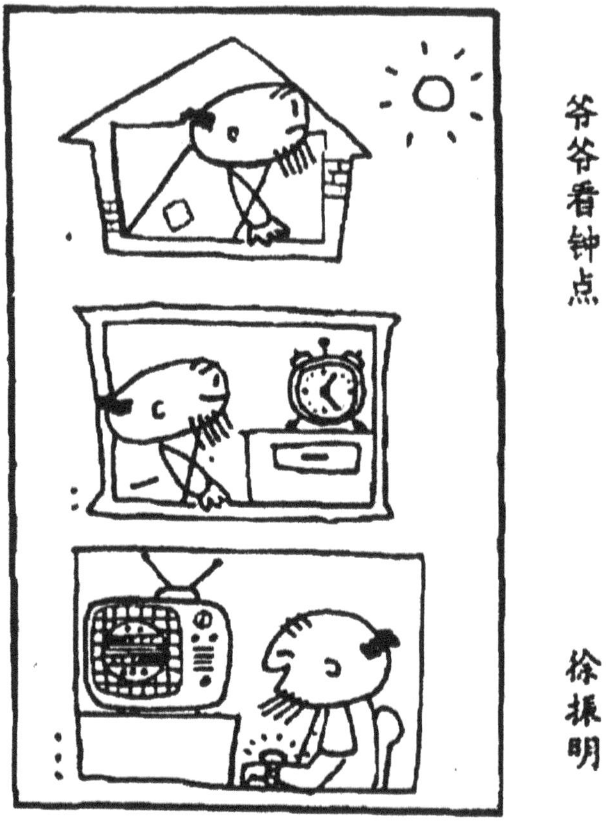

Figure 1.4: *Life out of 8 Hours*, 1982, no. 6.

The practice of giving a TV set as a gift arose from children's desire to bring happiness to their parents and fulfill their duty of filial piety, but no one could have predicted how the parents would use TV to fill in their empty lives. How seniors would occupy their time after retiring, while their children were busy with their own lives, became an important issue for both the individuals and society in the following years after the economic reform started. The relationship between filial piety and the ownership of a TV set has much more complicated cultural meaning as I have shown.

The Education of Children

Many visual materials also depicted children watching TV. However, in contrast to the images of older people watching TV alone, in most cartoons that showed children watching TV, the children were not shown alone but rather in the company of their fathers (see Figure 1.5 in Photo Spread). Lynn Spigel argues that in the age of the rise of television, the authority of the patriarchal father was eroded by his lack of knowledge about TV, and as a result the TV repairman replaced the father as the person most respected by the family. The fact that technology caused the average person to be dependent on an engineer was undeniable, and Spigel's argument about the decline of fathers' patriarchal power is also convincing. But the Chinese father's power to control and supervise in his family was not taken away; rather, it was strengthened, as is reflected in cartoons depicting children watching TV with their fathers.

A father's supervision was not necessary in the 1980s when the Chinese government strictly controlled TV content and TV stations could not possibly broadcast anything that was really able to threaten children's mental health. Nevertheless, images of children watching TV typically showed the father sitting next to the child. It is not easy to find cartoons from the 1980s in which a child is shown watching TV alone. Watching TV was presented as a family activity or at least an activity shared by father and child. In one cartoon, when a father and his son were excited by the soccer game shown on TV, the mother did not appear until they caused noise by falling off their chairs. (Gender roles in these depictions will be discussed in a later chapter.)

Rather than using TV as a babysitter, as Lynn Spigel argues that many parents did in the United States, Chinese parents treated TV as a teacher or a messenger to which they should listen *with* their children. The imperative of TV programming as officially ordered by the Chinese government in the 1980s was to construct a "clean" and "educational" content. Therefore, TV played a much more respectable role in the lives of young people than that of babysitter. It was not just company for children, as it was for older people, as the cartoons revealed; instead, it became a consultant and an educator, and the father is shown sitting next to the child as an interpreter or "classmate."

Both teachers and parents play important roles in the education of children in Chinese culture. Ma and Guo (1995) have recognized that for thousands of years, Chinese families have placed a high priority on their children's educations. Many classical works emphasize the role parents should take in educating children. In 1980s China, according to Ma and Guo, "schools and education institutions at all levels [were] invested with the task of turning out large numbers of competent and talented people suited to the demands of the modernization process. With such a heavy load on their shoulders, the schools [were] in urgent need of help from all circles of society" (1995, p. 6). Therefore, "parents school" was established as a way to "strengthen family and through it to enlarge the base of school-educated talented people" (ibid.). The TV set also brought education into a family, although the information it conveyed was not necessarily labeled as such.

As with TV in the United States in the 1950s and 1960s, Chinese TV also aired many documentaries in the 1980s. The documentary was one of the most important genres in constructing the role of TV as the presenter of facts. The visual images presented on TV made it a very persuasive educator. While scholars in the United States were struggling with the conflict between the needs of democracy and the risk of TV being used to spread propaganda, China was more concerned with how TV could fulfill its dual role as an educator and an entertainer. In the face of the Chinese government's clear statement that media should be the throat and tongue of the government and the party, nobody questioned the role of TV as one of the tools of propaganda in modern China.[1] Therefore, TV was much more than a babysitter.

In one advertisement (see Figure 1.6 in Photo Spread), a little boy was shown sitting in bed with two electronic toys in his hands. The antennas of the toys and the antennas of the TV are very prominent in the picture, and the mother- and teacher-like woman shown on the TV was Tongyun, a host of the well-known documentary, *Talking about the Changjiang River*. Babysitters might entertain or watch over children, but teachers and messengers needed to teach knowledge or convey new information to the audience. TV sets were playing a very important role in exposing children to a developing and increasingly scientific world. The advertisement says, "The TV set will accompany you in the process of reaching success." TV was depicted as more than a technological gadget or babysitter, and it became one of the many educators in the Chinese family.

Meanwhile, the role of TV as an educator or teacher did not take away the power of the father. Instead, the educator or messenger (TV) required an explainer and interpreter. The father's role as one of supervisor of moral education still existed, and the father had a new, welcome responsibility: accompanying his child while the child watched TV to receive new knowledge and information. This point was represented vividly in a cartoon titled "The Scary Lens," in which a father turned off the TV set in front of his son when there was an intimate scene (see Figure 1.7in Photo Spread).

Access to TV in China in the 1980s could sometimes provide much more knowledge than children could get in school. The Chinese Central TV (CCTV) and local TV stations all worked very hard to create programs that could educate children. As early as 1981, *The National Quiz Program for High School Students* was broadcast on CCTV. Later in the 1980s, there was a "Quiz Show Fever" in Shanghai (Guo, 1997). Parents who feared they would miss opportunities to improve the next generation felt pressure to watch these shows. Competition in attaining knowledge was a new anxiety for Chinese parents and children.

Elders watching TV alone while children watched TV with their parents also reflected the new pattern of Chinese family life in the 1980s. The young couples spent most of their time with their children instead of with their parents in the 1980s. This is also a mirror in which we can see the modernization of Chinese life. The most popular ways to practice filial piety and to educate children had changed from what they were in the earlier Chinese society.

The Replacement of Dialogue between Host and Guest

Among the few TV programs and dramas produced in 1980s China were two that themselves portrayed the role of TV in modern life. Scenes in both "*Zhong Gu Lou*" (The Bell Tower and the Drum Tower) and "*Xin Xing*" (The New Star) depicted the relationship between hosts and guests. *The Bell Tower and the Drum Tower* was one of the most well-known domestic TV dramas of the 1980s. It depicted the lives of a group of local Beijing residents. These characters' backgrounds varied widely. There were intellectuals of different ages, a Peking Opera artist, an official with a high governmental position, workers, and retired soldiers. The characters' different backgrounds contributed to a diversity of lifestyles and philosophies and to the conflicts that arose among them. They were all generally likable characters, but sometimes their different opinions about social issues and about life in general caused misunderstandings and friction in their communication and socializing.

In this TV show, a girl named Xing'er comes to Beijing from her home in a rural area to visit her father's old comrade-in-arms, one of the residents of the quadrangle. The old friends had once set up a betrothal between their son and daughter. However, the two families went on to have very different experiences during the next twenty years and their childrens' lives also diverged greatly. Xing'er, a beautiful and pure country girl, did not receive much formal education, whereas the boy, Li Zi, finished college and had just returned from England after studying there for a few years. Xing'er's father saved the life of Li Zi's father, Uncle Xun, during a war in the 1950s, and Uncle Xun felt ashamed to break the engagement he promised to his old friend. Both Xing'er and Uncle

Xun feel embarrassed when they think about the old engagement between the two families. One day, as they are watching a movie on TV together, the old uncle's guilt is reinforced when the sound of gunshots from the TV reminds him of the war. There is a generation, age, and culture gap between Xing'er and uncle Xun and there is nothing they can really talk about except the story of the two old comrades-in-arms and their promise. However, Xing'er asks a question about the war, and this exacerbates the guilt the uncle feels toward his old friend and the daughter. TV draws them into a conversation, and it connects them to the same topic—the war, the reality of the war, and the experience shared by Uncle Xun and Xing'er's father. In this episode, TV not only is a "helper" during the embarrassing and long silence, by providing something else for Uncle Xun and Xing'er to focus on and avoid starting a cold conversation, but it also creates a situation in which their hearts could be brought closer: by showing the "right drama."

Meanwhile, things shown on TV also connected contemporary life with an older historical moment and set up a complicated situation in which people of the past and of the present differed greatly. By doing so, TV expressed the powerlessness of people facing the significant social milieu caused by political, war, and economic issues. It was very hard in the 1980s for people to look to the past for models about how to navigate the world, which was very different politically and economically. In *The Bell Tower and the Drum Tower* Uncle Xun liked Xing'er and thought she could be the ideal daughter-in-law, and Xing'er liked Li Zi and his whole family. Although Uncle Xun and Xing'er had a special attachment, they could not help but accept the fact that they differed too much and could not possibly stick together. Xing'er was very reasonable, polite, and innocent, and she left with a smile and regards to the family. They both went back to their separate lives. Throughout the show, the only conversation between the old uncle and Xing'er was the one that took place while they were sitting in front of the TV. The conversation both reviewed the past and provided closure. The striking difference between the war on screen and the real lives of uncle Xun and Xing'er evoked nostalgia and revealed that the only connection between the two families was their personal history; however, their lives had to move on in completely different directions.

The Bell Tower and the Drum Tower was not the only TV show in the 1980s in which a climactic moment happened in front of a TV set. *The New Star* portrayed a young officer who had ambitious hopes that China's Economic Reform would allow him to make changes in an old city's outdated governmental policies. However, he was greeted with objections from other local officers and had to seek the support of a superior officer. The young officer went to the superior officer's house and wanted to get support. The superior officer did not care about the young officer's ideas and continued watching TV. The young officer keeps talking but the noise of the sports match the officer is watching drowns out the young officer's words; the superior officer responds only with a desultory "ah" or "en" from time to time. The young officer is very disappointed by the attitude of the superior officer, who uses TV as an excuse to avoid the conversation and the threat of confrontation between them. He cannot let the young officer leave without letting him have his say, but neither does he actually want to listen to the younger man; therefore, TV becomes a tool with which he pretends to listen and which helps prevent an argument.

The TV set was a prestigious commodity that the older man was given because he was a high-ranking provincial officer, for the purpose of educating him about the world and making him more aware of any changes in the nation. It is ironic, therefore, that the superior officer used it as a tool to avoid hearing about the young officer's ambitions and his plan to change the local situation and solve problems caused by the outdated economic system and stubborn colleagues. This ironic scenario reflected a real difficulty facing the Economic Reform movement in the early 1980s. TV could affect people's lives in many ways, but it could not really change people's thoughts as easily as the government supposed it could.

Many hosts depicted on TV shows were similarly depicted as relying on the TV for help during uncomfortable moments. Frequently, angry couples who were temporarily not speaking to each other were shown sitting together in front of their TV sets to show their existence in each other's lives. When guests and hosts did not know what to say to each other, they would sit in front of the TV and use it to resolve the embarrassment. TV shows provided a common topic for people to talk about, especially when they were from different backgrounds. The story in *The Bell Tower and the Drum Tower* was a typical one because Uncle Xun and Xing'er belonged to two different generations and lived in different cultural environments. Uncle Xun is an urban resident whose memories go back many years, and Xing'er is a young girl from a rural area. Without TV providing them with a common topic, there would have been no way for them to break the ice. Uncle Xun dares not mention the betrothal, and Xing'er does not know how to start the conversation. TV, which is outside the embarrassing situation, gives them common ground. They can relate when they watch the war being shown in the drama. The TV show was like an invisible presence that connected Xing'er and Uncle Xun. It initiated and participated in their conversation. The conversation that should have been started by host or guest was instead initiated by the TV. TV functioned like a third person, like a bystander to their conversation that helped them avoid a difficult and direct one-to-one dialogue. The real problem did not go away but Uncle Xun and Xing'er know that the process of sitting together, watching TV, and talking about something related to their thoughts is a way to resolve this communication problem.

Was TV helping or further blocking the communication between audiences? This question does not seem to arise in *The Bell Tower and the Drum Tower* and *The New Star*. The message seems to be that there are certain things that people do not have to talk about, and there are many things that will be better off if people pretend they do not exist at all. The two dramas show that TV provided a natural way for hosts and guests to avoid embarrassment or pretend not to notice embarrassing situations.

As people who lived in China in the 1980s know, when TV first became available, watching TV was usually a group activity, participated in with neighbors or co-workers. TV became a communal visual technology in both personal living spaces and public working places. It was not like TV watching in the United States in the 1950s, which allowed viewers to create a private space by setting up a TV room in which they sat with family members to watch (although TV did divide the family according to gender and personal interests in the programs). In contrast, in China, watching TV was a new, if not a "better," way in most Chinese people's eyes to maintain the community.

Did TV take the place of face-to-face communication or improve it? This was a complex question for most Chinese people in the 1980s. What could they talk about after enduring the chaotic Cultural Revolution and while facing the new political and economic environment? In the 1950s and 1960s, they were talking about the dream of the unification of the whole of China, the dream of having a society whose glorious spirit and social equality were more important than material concerns such as social status, power, and money. In the 1980s, what did people have to talk about? It was like the moment between Xing'er and Uncle Xun, which displays the change between generations. In their social life, change was a topic on which people could connect. Changes in governmental policy, in people's thoughts and lives, and in the scenery of cities and suburbs were the dominant topics in the media of the 1980s. TV itself gained its popularity because of these changes; at the same time, it displayed and disseminated a message about the change to the Chinese audience. Change became the main topic between the host and guest, among the family members and neighbors.

The Relationship between Materialization and Intellectualization

The change brought by TV and represented by TV also became a controversy in intellectual arguments, as was reflected in the comics and in articles in magazines, journals, and even the early TV dramas. What is the meaning of TV?

What does it represent in Chinese culture? How did it connect to people's beliefs and daily lives at the transitional moment between the Cultural Revolution and the Economic Reform in the 1980s? Looking through various representations of the relationship between TV and people's lives, it is not hard to see that representations of TV exemplified conflict between modern, developed culture and traditional values and moral standards.

In many visual materials, criticism of people's morality was expressed by showing their desire to own a TV set and to watch TV instead of doing more meaningful things, such as studying or taking care of their parents or family. However, at the same time, there also were many representations of TV that praised its benefits, such as its capacity to educate.

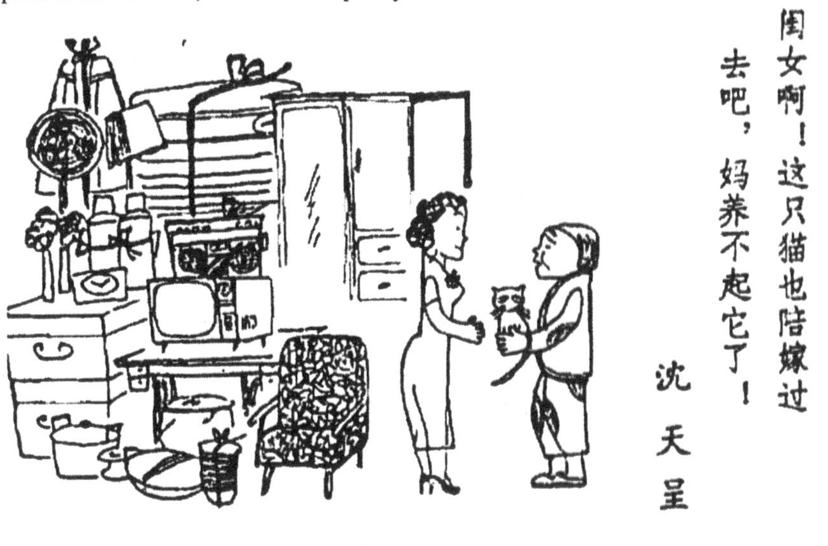

Figure 1.8: *Life out of 8 Hours*, 1981, no. 4. In this cartoon, the old mother is telling her daughter, who is getting ready to marry, "Girl, could you take this cat too? I cannot afford it any longer." The cartoon is criticizing the fact that in order to have decent weddings and display material wealth, young people took with them too much of their parents' money. A TV set is the centerpiece of all the wedding gifts from parents to the children.

As a means for the transmission of knowledge, TV not only aired formal education programs but also introduced to the audience the fascinating world beyond China, including different lifestyles from around the globe. Many people had learned about the food, drinks, homes, transportation, and customs of people in other parts of China and other areas in the world through watching TV. The TV set represented Chinese people's abandonment of morality, tradition, and spiritual pursuits, but its educational role was portrayed as positive and significant, capable of enriching people's knowledge and opening the audiences' minds and bringing them closer to the world that existed beyond their own daily lives (see Figure 1.9). The Chinese government was very confident that it was successfully controlling the content that could be accessed by people through TV. Chinese elites worried about the consequences of surrounding people with images of material goods, instead of higher morality and spirit, but the common people realized the advantage that material goods could bring. The conflict between the two ideas embedded in the depictions of TV illustrated that Chinese commercial culture was still immature. In the 1980s, Chinese media had not started to play the role of promoter of commodities, and accepting that commodities were necessary was still very hard for Chinese people. It took time for most Chinese people to switch from buying to fulfill needs to buying for pleasure in the early years of Economic Reform.

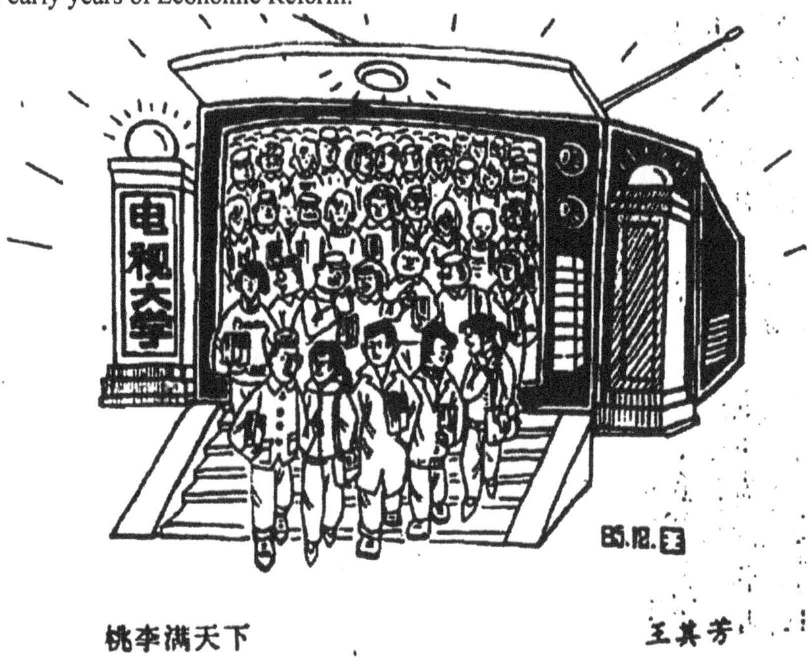

Figure 1.9: *Life out of 8 Hours*, 1986, no. 3. Titled "With Pupils All Over the World," this cartoon portrays hundreds of people "walking" out of a TV set, which represents a TV university.

The binary roles of TV as a materialized commodity and as a carrier of a spiritualized modern ideology echoed the goals of Chinese Economic Reform. The representation of TV reflected the government's statement that economic reforms were aimed at helping people gain a higher level life, but at the same time, it needed to follow the highly ideal motif of the government. The Economic Reform was a necessary step for Chinese people to pursue a better life. TV, therefore, became a representative technology rich in symbolism in the 1980s. The symbolic meanings of TV cannot be compared to those of other products, such as the refrigerator, washing machine, or the stereo set, even though together they were categorized as the "New Four Big Domestic Items" at that time. The other three "big items" do not have the same characteristics as the TV. Using them does not require people to stop their usual activity or housework; instead, the items make doing traditional housework easier without conflicting with other activities, and they make daily life much more convenient. For example, people do not have to cook every meal when they can keep leftover food in the refrigerator. However, using TV requires people to make time for it. People need to rearrange their daily routines when they are at home if they want to watch TV because of the broadcasting schedule and the fact that people cannot really do other housework if they want to really "watch" TV, that is, watch it without interruption. The meaningful content of TV in the 1980s conveyed the message that TV was not just a killer of time and that it was not to be watched randomly and carelessly. According to Rick Altman (1986), many people in the United States think they are watching TV but are actually only listening to it. Sound draws people's attention to TV programs. However, Chinese people were excited by this visual and auditory machine. They were engaged audiences in the 1980s. Because the state controlled TV programming, viewing was regarded as legitimate. It was a window into a much bigger world than Chinese audiences had known; it was like a class or textbook providing knowledge and conveying news and governmental policies to people. TV was important. Both the human agencies and people were making TV meaningful in their life.

Magazines and newspapers in the 1980s represented TV as exhibiting a foggy boundary between entertainment and education. This reflects the dilemma between the goals of promoting economic development and of sustaining the pursuit of a society in which political ideology overrides everything else. The spirit sought after dictates that people should be equal in owning substantial things. However, it is very easy to identify a satirical tone in comics that criticized the unequal ways in which people spent their time when they were off work. The officials were thinking about entertaining themselves and looking forward to their personal lives; however, the young intellectuals could not help but keep worrying about how to finish the housework and help the family when they had to devote themselves to their work all day long (see Figure 1.11 in Photo Spread). At the same time, TV was also the main piece of "evidence" when cartoons criticized officials' corrupt behavior (see Figure 1.12 and Figure 1.10 in Photo Spread). This was because TV sets were precious and expensive and people bought them as "gifts" for officials in exchange for favors. At the same time, TV used up time, and one of the most important moral attributes expected of a good official was that he or she devoted his or her time to serving people and helping people solve their problems.

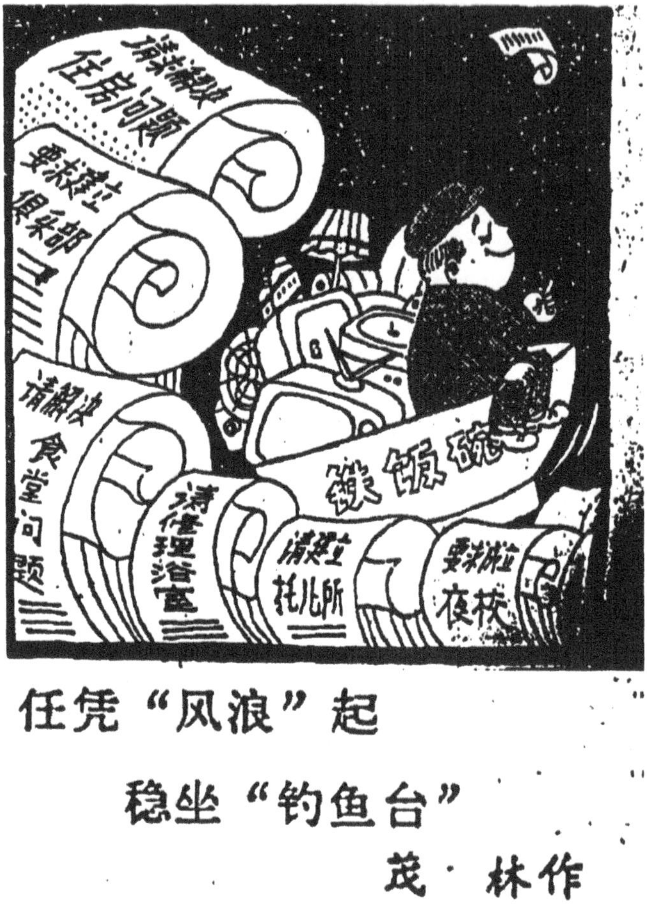

Figure 1.12: *Life out of 8 Hours*, 1980, no. 6. This is a satire of an official who does not take care of people's applications but enjoys his privileges. A TV set is next to the official.

This potential harm to a virtuous officer was depicted in *The New Star*, one of the earliest TV dramas promoting the Economic Reform (Guo, 1997), when the superior officer avoided listening to advice he should have taken seriously. In the cartoon (see Figure 1.11 in Photo Spread), the comparison between a superior officer and a hard-working intellectual also exposed what I have just argued above that TV was used as a piece of evidence to criticize the dysfunctional roles of some officials. Therefore, the representation of TV sets in popular media symbolized people's dark sides such as greediness, laziness, irresponsibility, and selfishness. What terrible symbolic meanings TV incurred when it had just appeared in Chinese people's lives! It was the representative figure of modern technology and it was the substance opposed to the ideal of highly spiritual and moral pursuits. Meanwhile, how wonderful to have at home a TV that could entertain family members and educate children! The two distinct tones used when speaking of people's relationship with TV in a "new" and changing society created as well as reflected confusion, a confusion about modernization and human beings. TV is controversial among the critics of modernization in magazines and newspapers, but it also is inspiring and fascinating to Chinese people. Chinese intellectuals are always worried about the good and bad part of pursuing a highly modernized society imitating the foreign models.

Chinese people were struggling between the desire to not own an unnecessary commodity and the undeniable fact that they needed a better way to entertain themselves than listening to radio.

Mitchell Stephens (1998) has pointed out that one of the lessons of the "Communications Revolution" is that new media are usually attacked while they are young. The attack on television in its early stages in China is an example (see Figure 1.13 in Photo Spread). Except for some harmless criticism of the dramatic narration of TV dramas (which I will discuss later), attacks on television in China centered on the ownership of the TV set and the TV set's reflection of traditional morality during its early "immigrant" stage. The symbolic meaning of television is much more important than its scientific or informational functions. We all know that if a TV set lacked the capability to transmit information and news through visual images and sounds, it would be nothing more than a piece of useless furniture. However, the criticism of television expressed in the visual records in the magazines in the 1980s paid little attention to what television, as a media technology, could bring into Chinese society but focused on what television, a lifeless item, represented.

On the other hand, the cultural and ideological content of TV programs was glamorized by the Chinese media. This interesting phenomenon—the contrast between the media's depiction of TV sets and the media's depiction of TV programming—can be much easier to understand if we notice that the physical body of TV sets was a product of commercial/capital culture; however, TV programming was selected by the socialist Chinese government. The criticism of the "body" of TV warns of the incoming and new outside culture it brought, whereas the praising of the information carried by TV announces the safe and healthy content that socialist China had promised. Although by the 1980s, TVs were being manufactured within China and the TV industry was a source of national pride, the dominant Chinese intellectuals still thought, mostly unconsciously, that TV sets were a symbol of commercial production that people should be careful about. The typical intellectual had not realized that the technology itself could create a change that worked in the opposite direction of the manipulated information or ideological content it was made to carry. For most critics at that time, a potential negative influence of television was its creation in people of a fondness for consumer goods and an abandonment of the regular pursuit of a spiritual life. Intellectuals feared that people would become materialistic and care too much about the things they owned, and forget that some things were much more important than whether they owned an expensive item or not, or whether they were modern enough or not.

Media representations of TV convey Chinese intellectuals' original and basic fears of commercial culture. However, the media representations also reflect ambiguity among intellectuals as to what commercial culture actually was. Why did the media focus on television rather than gold or silver, the most precious substances, to express the fear and the worry and to send the warning? They did so because television combined newness, modernity, informational technology, exoticness, and also openness. These characteristics, novel to Chinese culture, provoked people's curiosity and seemed to offer a path to greater prosperity, a goal of many Chinese people in the 1980s.

Conclusion

There are various dimensions to account for when examining the relationship between tradition and modernity. It is hard to draw a line between any of them, as we can see from the case of the representations of the TV set in Chinese media in the 1980s. The value system is embedded in and, to some degree, decided by the other dimensions, such as political ideology and economic policies. Actually, there isn't a clear line in between any two of the factors that we can think of in the process of modernization. Modernization, to some degree, is the disruption of tradition. The comparison between modernization and tradition is in fact not between two different concepts, but a constant process of the changing of tradition since the invention of modern technology. Modernization does not simply mean newness; it signifies a difference from the common perception of things, and from ideology and common practices at the moment that tradition is affected.

Although Chinese modernization in the 1980s consisted mostly of the adoption of Western technologies and business management practices, there was a necessary cooperation and negotiation between the imported newness and the domestic national ideology.

The dominant ideology of China in the 1980s was the state-sponsored Four Modernizations, in the areas of agriculture, industry, national defense, and science and technology.[2] The national development policy created new values and changed the traditional values at the same time. While Western culture brought in a great respect for modern technology, the relevant traditional issues that have always been central to Chinese culture were also reformed. The representation of TV sets in the media is one of the great examples of the revision of tradition at a particular historical moment.

The visual representation of TV sets in 1980s Chinese magazines, newspapers, and TV shows exemplifies a complex of discourses caught between the power of authority and popular resistance, local and international ideology, and historical and current culture. Chinese society was undergoing a transformation from a purely socialist economy in which the dominant political system controlled everything to an economy in which the government attempted to give the "market" some power to operate and direct production and stimulate people's desire to consume. However, the government and the media were very careful about where the new policy would lead the culture and society. The Chinese media were using the old themes that people cared about and at the same time were apparently harmless to the dominant ideology. To some degree, the media's representation of TV reflected the conflicting emotions of China's dominant ideology in the 1980s; people, represented by the voice of the media, were happy about the energy created by the Economic Reform in motivating a prosperous market and a much more vital culture but were worried about the problems caused by the conflict between the national and the foreign ideology, especially the divergence between increasing materialism and the lofty pursuit of spiritual enrichment. The replacement of dialogue between the host and guest portrayed in the TV dramas displayed the role and function of TV in a newer cultural environment. At the same time, it also illustrated the increasing gap between people of different generations—not only the common people like Xing'er and her Uncle Xun, but also the young official and the old official. Television could create conversation to bridge the gap between generations, as well as take away the conversation when one party does not want to talk. However, TV decreases the awkwardness that usually happens in either situation. Is TV important in an evolving society? In the various representations of media, there were imagined solutions to people's needs, and also the promise that magazines, newspapers, and TV shows made for TV. It was a discourse full of both joy and worry.

Notes

1. In his speech "The Only Way to Unite is to Depend on Ideals and Disciplines," presented at The National Conference of Science and Technology (1986), Deng Xiaoping expressed again that media should pay attention to their role as the servant of governmental policies and ideology (*China Daily*, March 9, 1985). Hu Yaobang, the General Secretary of the Communist Party of China Central Committee in 1985, also emphasized the same point in his speech, "The Party's News Policies." Hu declared clearly that all the news institutions should serve as the tongue and throat of the central government (Chinese Yearbook of Broadcasting and Publications, 1986, pp. 5-17).

2. These were first proposed by Mao Zedong in 1964. Deng Xiaoping made the plan more realizable by designing detailed quantitative goals in 1979.

CHAPTER 2
JUMP STARTING A LEARNING PROCESS: THE OWNERSHIP OF TV SETS AND SOCIAL STATUS

Many new inventions serve as symbols of wealth for people who can afford them until they trickle down to the rest of society. Even ice cream was first tasted by the Chinese emperor. The appearance of television in the United States in the late 1940s exemplifies this trend. Most people who could buy a TV set belonged to a middle- or upper-class family. Contrary to merchants' expectations, television was not immediately welcomed by a large group of consumers, although sales did take off throughout the 1950s.

In China in the 1950s, TV was beyond the means of most people. Only an extremely small group of people, made up of governmental officials, owned TV sets, which were given to them by the government. Television sets were not a commodity in the 1950s in China but a privileged gift, owned by only a few "special" people. Even for these people, the TV set was more like a piece of expensive furniture because the broadcasting system was in its nascent stages and there were very few programs to watch. Television, and the social status of TV set owners, began to be popular topics in Chinese mass media during the 1980s when TV became a real product and commodity.

Many Chinese people probably still remember the short skit called "The Blind Date of a Lazy Man," aired as part of the 1989 Spring Festival Gala held on CCTV. This comedy made many viewers laugh so hard that they cried. It describes Panfu, a lazy man who, unlike others in his village, does not like to work to improve his life and is still unmarried at age forty. The village head's job was affected by Panfu's laziness because the village could not become the model village with Panfu being poor and unaccomplished. The village head decides to help Panfu find a wife, thinking that being married might encourage Panfu to work harder and improve his life economically. Because Panfu does not own anything at home except a thermos bottle, the village head gathers some balloons, a sofa slipcover, and a TV slipcover and carries them to Panfu's home to help him "decorate" his barren home. Wei Shufen, the woman to whom the village head wants to introduce Panfu, cannot see well. The village head tells Panfu to put a balloon on a chair and then cover it with the sofa slipcover to simulate a sofa. He then puts a TV slipcover on the shoe box to simulate a TV. Panfu's home suddenly has "everything" that could demonstrate a wealthy standard of living, as the village head declares. The lazy man has spent only 26 cents on the balloon and a free shoe box, but the décor is great.

Viewers were expected to laugh about the drama that ensued when Wei Shufen came to the house in obedience to the village head's instructions. The audience can tell that the village head does not want to cheat the couple. His purpose instead is to help the two "old" single people get together and build a family; the head expects that Panfu will then be motivated to work hard and buy a real TV set and sofa after he and Wei Shufen get married. The drama ends with the plan falling apart after Wei Shufen discovers the faked sofa and TV set. She leaves, saying, "I will come back when you become a real good man." The village head's plan has failed. Viewers laughed when the village head and Panfu attempt to prevent disclosure of their faked TV and sofa and also at the crude accents and behaviors of the village girl. It was meant to be funny. Owning a TV set and a sofa did not necessarily mean the owner was wealthy; however, ownership of such items signified a comfortable standard of living. It was at least clear that the person did not have a food or shelter problem if he could afford to have a TV set at home, and so TV ownership became a condition demanded by most prospective brides and their parents. Ownership of a TV seemed to foretell a fine life and a firm basis from which to start a family. Few Chinese audiences questioned the ideology behind the short skit; partly because it was a comedy, partly because the idea behind it was widely accepted. How could a person get married if he could not at least have a TV set at his home? It was the easiest if not the best way to prove a person's ability to gain a good standard of life in rural China at the end of the 1980s.

Careful analysis reveals that this short skit is full of cultural meaning, and two points are particularly significant. The short skit shows that by the late 1980s, TV came to be considered a necessary electronic possession in many Chinese families, even in the rural areas, and watching TV was not as unusual as it was at the beginning of the 1980s. While the skit makes clear that ownership of a TV set was very common in the life of the village people, one should not ignore the evolution of the TV set before it became popular. Why was owning a TV set and a sofa thought to be so representative of a man's capability of raising a family? What did the cultural environment contribute to the symbolic meanings of TV in 1980s China? This chapter discusses the relationship between TV and the cultural context in the 1980s.

The Invisible Wire behind TV Sets

Most television viewers in the twenty-first century think the televised images appear at the push of a button, and leave it at that. Once we become accustomed to a technology, we do not think about how the technology works. But when asked to think about it, viewers might be surprised by the complexity of issues connected with the screen and by the number of steps that must precede the moment when we simply push the button. Sometimes viewers simply forget all these issues because TV is already part of our daily lives and we need not bother to think about them.

It is useful to consider what has made TV watching possible in people's daily lives. According to David E. Nye (1990), in the United States in the 1930s, the radical power of electricity and the success in embedding electricity in modern life pushed many corporations to explore manufacturing and selling new electronic appliances to extend the market. "A form of visual radio called television" was among all of the "new desires" (Nye, 1990, p. 360). Electricity is an invisible power, which can intimidate because of the huge machines and the complicated wiring that are mostly hidden from view. Electricity, as David Nye (1990) hints, changes the past, creates the present, and promises the future. Modernization rushed to a new stage once electricity began to be used in every area of people's lives. Even though scientists have found more powerful and useful sources of energy, electricity is still the primary engine of modern life. (For example, consider the results when electricity goes out in a big city like New York City.) Cities would not be at all pleasant without electricity. Electricity lightens the night, facilitates production of industry, and extends the hours in which work can be done.

China has had electricity in large cities, such as Shanghai, since the early twentieth century. The People's Republic of China's government devoted a lot of time, resources, and labor to develop technology, equipment, and experts who could increase the production of electricity. Gezhouba Dam, China's biggest water construction project of the mid-1980s, is an example of the effort that China has made to establish a foundation for a faster modernization process. (I will continue to give more analysis on this point in Chapter 3 when I examine the relationship between modernization and nature.) In short, as China sped into modernization, the nation's electric-generating capacity increased, lighting the night in the countryside and distant towns and cities where the supply of electricity had been very unstable before. Zhang, an interviewee from northeastern China, said, "I bought a candle factory in the early 1980s because I thought it would bring me a big profit since the electricity often went out at that time, but unfortunately, as soon as I bought the candle factory, we seldom had a night in which we did not have electricity. The supply of electricity in our town was very steady. So I lost lots of money because of this unwise investment. You know, electricity at least went out for an hour every day during almost half a year around 1982 and 1983." Zhang was an elementary school teacher in the suburb of Da Lian before he decided to become a business owner, but apparently, it was not a successful experience.

Zhang's story reflects the speedy development of electricity in China in the 1980s. Typically, it took less than half a year for a suburb to acquire a constant and steady supply of electricity. Therefore, it is not surprising that the purchase of domestic electronic appliances increased dramatically during the early 1980s. The "four big" appliances, the refrigerator, color TV, stereo, and washing machine, are all electronic appliances. Two of them, the refrigerator and washing machine, are time-saving appliances that allow people to go grocery shopping less often, to not have to cook every meal from scratch, and to do laundry less often; the other two provide entertainment. But as noted before, the entertainment that the stereo set provides is different from what the TV set does, because the auditory media do not require people's visual attention. For the excited and devoted Chinese audiences in the 1980s, TV sets required much more attention. Modern machines seemed to promise much more time to people, especially housewives who were still taking care of the largest percentage of the housework at that time, time that women might use, at least in the United States in the 1950s, to watch TV with their families. Electricity not only illuminates the night and extends the nonsleeping hours of people's daily lives but also, by liberating people from many daily chores, gives people more free time in which to entertain themselves. A steady flow of electricity is the technological foundation for the continued modernization of people's daily lives. Although the people who were enjoying the benefit of these electronic appliances in early 1980s China were still few in number, electricity, technology, and the promotion of the idea of modernization promised all people a better future (see Figure 2.1 in Photo Spread). However, the more free time that most people got in the early 1980s was not due to the promise of technology but due to the policies and situation caused by various human agencies.

Ms. Wang, an interviewee now in her sixties, said, "We used to do lots of things after we finished our work in our work unit (Dan Wei). But we were not as busy as our parents used to be in the 1980s because we did not have as many children as they had, due to the birth control policy. We did not have to cook as much as our parents had to and did not have to take care of every kid's issues every night. Things that we needed to finish up were not as a pile of things that were pieced together by everybody, and that was a big family with many issues. So we had time to watch TV; it was not like listening to radio while still having to work for the same amount of time we used to." As Wang suggests, if people of her generation still had as many children as did people in her parents' generation, they would not have had time to watch television even if they owned a TV set. Technology did not by itself shape the schedule of people's lives in the process of modernization. Technology might make people's work less tedious, but people still need to work a certain number of hours every day. Although technology could save people more time from housework, not many people chose to buy a washing machine or refrigerator before buying a TV set in the early 1980s. Most families, paradoxically, bought the TV set before they bought any other domestic electronic appliances. It shows the power of desire and imagination in shaping the meaning of technology in people's lives. And many Chinese families, especially rural families, did not buy their first refrigerator or washing machine until the 1990s. In general, Chinese people were much more eager to own a TV set than to own one of the other major electronic appliances. A TV set "improved" people's lives in a crucial way.

From the beginning of the People's Republic of China (P.R.C.), people's non-work hours were mostly occupied by attending various political meetings and conferences and by doing housework. People did not have the option to not attend the meetings and conferences. During the Cultural Revolution, such political meetings and conferences were the main activity of people's daily lives besides the work they had to do in developing the industry and agriculture of the nation. Kevin Latham notes, "Political education and meetings in off hours would be conducted at the work unit—in communal halls, on sport fields, or in factory yards, for instance" (Latham, 2007, p. 236). People did not really have that much time to engage in personal activities. Women had to take care of a big, often huge, family. Many places did not yet have electricity, which meant usable evening hours at home were too few to spend on leisure activities. People in rural areas did not even have time to get a full night's sleep, after cooking a big meal, doing dishes, washing clothes, feeding pigs and chickens, and then having to get up very early to start the cycle again. The ability to listen to a radio while doing housework was luxury enough for most people. Therefore, suddenly, after 1979, and especially in the 1980s, because of the one child policy, and the economic reform, people did not have to take care of many kids or participate in the political education and meetings anymore. Therefore, how to make evenings more interesting became a big issue. As Latham (2007) argues, "leisure time was depoliticized in the 1980s," and "shift[ed] from being collectively oriented to being individual and family-oriented" (p. 235).

As a consequence, TV became the most important appliance in many people's lives. Young people could learn from the Television Broadcasting University after work, old people could watch TV programs, and middle-aged people could relax a little bit after they finished their long day. Caring for the needs of a small family required much less time than taking care of a big one.

For newly married couples, the Chinese government's birth control policy prescribed that their future family would be a small, three-person family. One of my correspondents reflected on why she thought it was very important for a young couple to buy a TV set when they got married in the 1980s: "The man should at least buy a TV set as the betrothal of the marriage; otherwise, what does he want his wife to do with him at night? Just turn off the lights and produce a baby? We didn't even have to produce that many babies anymore. So the young couple, especially the wife, strongly required a TV set when they were talking about marriage. People could wait and buy other items later, but a TV set, it was so necessary!"

The passionate desire to own a TV set, although it symbolized people's relative wealth in the 1980s, paradoxically also reflected the shortage of other leisure activities at the time. Owning a TV set actually provided an inexpensive entertainment, because people did not have to pay extra money, except the fee for electricity, on a regular basis for this simple activity. And they were able to figure out ways to save on electricity. Many people watched TV without turning on household lights not only because the darkness increased the visual quality of the TV but also to save money by conserving electricity.

Learning the background in which people bought or looked forward to buying a TV set in the early 1980s, we cannot but stop and question the symbolic meaning of TV sets created by the Chinese media in the 1980s. While most people needed a TV set because of the desire created by the social context, the display of TV sets in most people's view and memory became prestigious. The symbolic meaning of TV sets was not just based on their economic value but also on other ideological issues that can be seen from the visual representation of TV sets in the 1980s Chinese media. An examination of how the media depicted television and TV sets could help find out the symbolic meaning that was given to TV sets in the 1980s, an important phrase for China's transformation from a communist regime to a regime focusing on economic development.

Having a TV Set and Being a Social Elite, or Getting Better off Earlier

In the 1980s, Chinese mass media frequently presented images of famous people standing or sitting next to a TV set. In one picture (see Figure 2.2 in Photo Spread), historian and professor Zhou Gucheng, wearing a pair of glasses with dark lenses, is facing the camera and smiling. Aside from the professor, the focal point of the picture is a TV set in the middle of the photo. The picture shows that the historian's ownership of a TV set is as important as the portrait of his intellectual life with Chinese calligraphy and history books. Pictures (*Chinese Pictorial,* 1986, p. 5) depicting the happy life of a Taiwanese couple after they decided to move back to mainland China in the 1950s were shot by the photographer with the entire family gathered around the sofa, smiling, with a TV set fully visible behind them. Apparently, the people on the sofa had the TV set behind their backs. In pictures of American families, the sofa almost always faced the TV set, and the living room would be arranged with the TV set as the center. However, in the pictures displaying a TV set in many Chinese magazines, usually both the TV and the sofa were facing the camera. In contrast, if we look at the photos published in Chinese journals in the 1980s, we may find some pictures in which a person is smiling at the reader with a computer facing the reader at the same time. This representation of the division between people and technology often has a symbolic meaning.

In Spigel's examination of TV sets in the United States in the 1950s, she discovered that pictures usually showed people facing the TV set, watching the programs, and with their backs to the camera. However, in Chinese magazine photos, often both the TV owner and the TV set are facing the camera and the TV has a blank screen. It is as if the TV is not for the person to watch but both the TV set and the person are for the readers to watch. The representation of TV in this style is significant, because putting a TV set behind the owner and especially the sofa is very awkward and deliberate. It was considered unnecessary to turn on the TV set in the portrait of TV sets. These media images implied that television affects society even when TV sets are not turned on (see Figure 2.2, 2.3, 2.4 and 2.5 in Photo Spread).

An even more fascinating picture is one showing the brother of the last Chinese emperor, Pujie, smiling at the readers, with a TV set sitting behind him. This is a dramatic juxtaposition of the traditional nature of the dynasty (mysterious and stubborn) and modern media technology. The photo seems to imply that the past has gone, the future is waiting, and the TV set, a modern technological product, is a bridge between them. Published photos of this type are no longer common. Images in Chinese media since the 1990s very rarely depict social elites sitting next to a TV set. In the twenty-first century, professors are portrayed in a way that makes more sense to us—usually sitting in front of a full bookshelf. But in the 1980s, it was a different story.

Chinese mass media in the 1980s also frequently showed pictures of peasants with TV sets (see Figure 2.6). Such pictures often included other electronic appliances, too. That these individuals were rural residents is very easy to discern because of their peasant headscarves and items such as cattle, poultry and farm products. These images suggested that peasants' hard work and wisdom in following the party's economic policy brought them better, more prosperous lives.

Many pictures in which a TV set appears have three common characteristics: nothing is shown on the TV; all human faces, along with the screen of the TV set, are facing the readers; and the person at the focal point is often culturally or politically special. TV's role as a transmitter of programming did not seem that important in China in the 1980s. The photographers were more concerned with showing ownership of a TV set and making sure that the TV set was visible. The appearance of TV sets in the visual portraits of people's lives was too common to be questioned by Chinese readers in the 1980s. The smiling faces of the people in those pictures echo the general opinion among Chinese people in the early 1980s that life would be resurrected and the economy would recover.

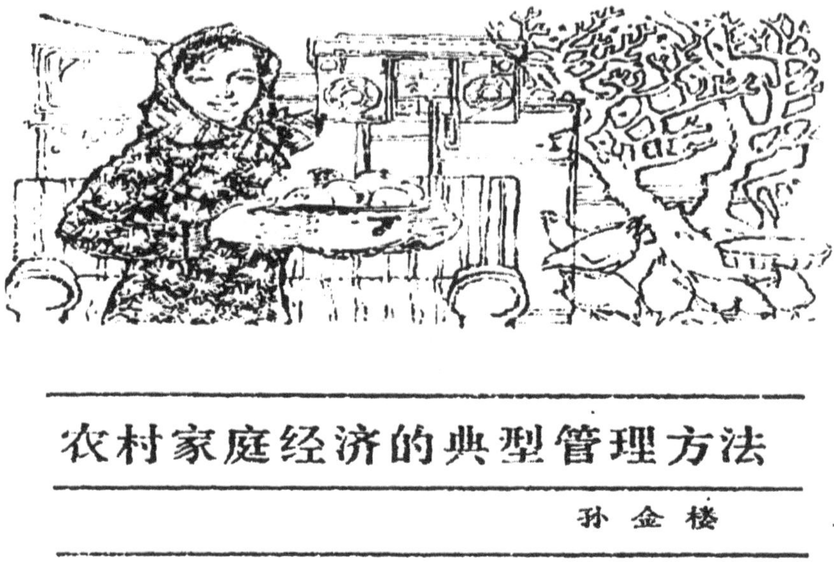
农村家庭经济的典型管理方法

孙金楼

Figure 2.6: *Modern Family*, 1985, no. 2. This is a village woman who is getting rich. There is a boom box on her right and a TV set on the left.

Owning a TV set meant more than having access to entertainment. As discussed above, a TV set also signified social status, the governmental policy of providing benefits to important people, and a kind and merciful attitude toward people who, like the last emperor, might be in danger during the transition from the old political power to the new. In the picture of the brother of the last Chinese emperor, he appears happy and peaceful as he sits with a TV set behind him and with ink and a Chinese brush on the table in front of him. The photo portrays a "reformed" feudal ruler devoting himself to his intellectual life by reading, learning, writing, and engaging in retrospection. While he is grasping the traditional writing tools, which were tools of an art form rather than of everyday writing, his life has also been enriched by television. Although the government had declared the emperor to be a common citizen, it had always given him special care. His material needs were completely taken care of, thus allowing him a relatively high standard of living.

In most of the photos of famous people with TV sets, viewers will not see anything precious or expensive aside from the big screen TV set, and perhaps a sofa, which makes the room look nicer and warmer. The TV set thus often became the focal point in such representations of people's lives, implying that a TV set is a symbol of a better life and of an intellectual life. The TV set competed in the photos with the traditional calligraphy and Chinese water color as the central decoration of the home. Whether the TV set was actually more important than other things in these people's lives, the photographers thought the TV was the most important thing. While American media were trying very hard to create images of a harmonious, family-centered environment for the television watching experience, Chinese media were deliberately connecting television with the social status of social elites. The existence of a TV set in an elite's life was naturalized in these Chinese media representations in the 1980s.

Advertising Television but Not Advertising "the" TV

Parallel to the representation of the TV sets in the lives of these social elites were representations in other TV advertisements of lives that most Chinese people would not have a chance to experience. In the picture of this Golden Star Color TV (see Figure 2.7 in Photo Spread), some Europeans are stepping out from the TV set and the European landscape it depicts. As shown in this photograph, TV sets were bringing images of non-Chinese places and people into Chinese people's lives. The commercial's slogan, "Don't you feel happy when your friends are visiting you from a faraway place?" is a quotation from Confucius' *Lun Yu* and is one of the philosopher's best-known statements. The exotic images and the familiar, traditional idiom are harmoniously combined under the themes of friendship, distance, and newness. The advertisement conveyed the notion that having a TV set made it possible for a person to know an "unknown" outside world without going there; rather the outside world would come and visit one through one's TV. This ad does not make an effort to stress the uniqueness or merit of the brand, as ads usually do; instead it emphasizes that owning any TV set is necessary for the purpose of learning more about the outside world, and that TV ownership is a way to bring new and interesting friends into the viewers' lives. Television was treated as a new invention, and the ads needed to first tell people why it was necessary to own a TV set: doing so would be like welcoming a friend, who would bring newness and exoticness into people's lives. The combination of the ancient idiom and the new electric technology "naturalized" the role that TV could play in Chinese people's lives. It is just like a friend, and one should be hospitable and welcome its coming.

The ad portrays TV as a messenger or liaison that could lead Chinese people closer to unknown cultures and habits. Apparently the advertisers assumed this unknown life and culture to be tempting for the potential magazine audiences. The idiom blurs the relationship between having a modernized life and learning from other countries. Among the advertising photos depicting modern families is one that shows young parents with their son and daughter sitting on a sofa, in a clean and warm modern home, apparently watching TV. The family members' faces resemble those of Chinese people, but the clothing, the hairstyles, and especially the fact that there are two children all denote that this is not a typical Chinese family. Due to the popular belief that Japanese people had more modern and generally better lives, it would be likely to be taken as a Japanese family. Most striking is that there are two children, a deviation from the Chinese government's one-child policy. Therefore, the audiences are encouraged to peek into a foreign but culturally similar family. In the corner of the picture is a set of stereos, which reinforces the ideal that the home is modernized.

The ads present a potential future for Chinese people. The home is clean, neat, and comfortable, and the family members appear to enjoy spending time with each other.

Television is a symbol of a new modernized lifestyle. Magazine advertisements hinted that it could both teach about and enable a new lifestyle. The declaration on advertisements' role in China in *Wenhui Bao* (a newspaper published in Shanghai) makes clear the deeper social meaning of the portrayal of TV sets in the media.

> It's necessary to take advertising as a way of promoting the international trade, and improving the business management. We should use advertising as a channel through which to provide knowledge and make it convenient for people to communicate and establish a closer relationship between the producer, retailer and the masses. (Xu, 2006, p. 231)

This statement suggested that the role of advertisements about TV was not to stimulate a desire for TV as a commodity but to provide a service to society in general.

The periodical press connected the TV set with images of modernized homes. However, most people who had a TV set could not even afford a real sofa, which was supposed to be part of the viewing experience. In images published in 1980s media, people were often shown sitting on a little wooden bench. The TV experience did not necessarily include being physically comfortable; just being able to view was sufficiently satisfying, in the cartoons' depictions. This aspect of how TV became part of people's daily practice in the Third World has been discussed thoroughly by scholars.

John Tomlinson's photo (*Cultural Imperialism,* 2002, p. 1), showing an African family watching television in the desert without a chair, is an iconic image. If in pre-modern society, the ironic phenomenon is that a family in extreme poverty still has decent clothing for the male leader, the equivalent in modern society is the ownership of a TV set by an otherwise destitute family. While advertisements, such as that shown in Figure 2.8 in Photo Spread, were setting up an idealized home scene in which all the furnishings match, the reality was that many Chinese people in the 1980s were watching a small 12-inch or 14-inch TV while standing amid a large group of people. People watched the TV programs without room and space for a bench or chair, not to mention a sofa (see Figure 2.9). However, the advertisements tended to portray viewing practices of people in developed countries and thus created a model of the most modern way to watch television. Therefore, modern home décor should include a TV stand, a sofa, and wallpaper, all of which most Chinese families still did not have in the 1980s.

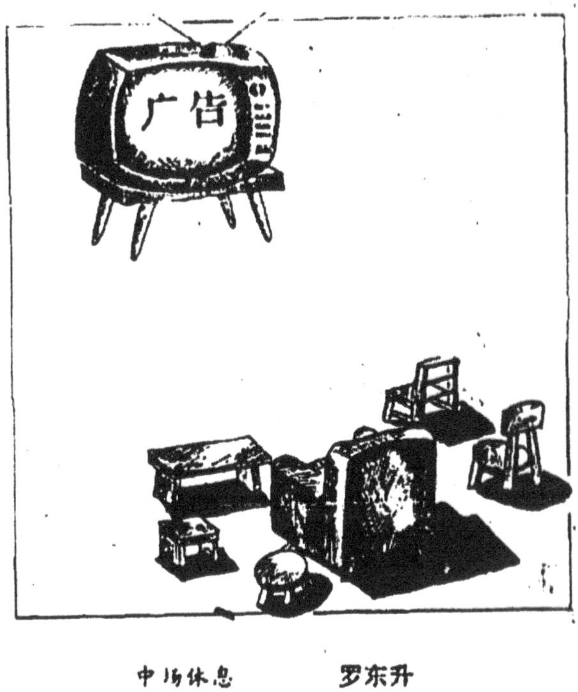

Figure 2.9: *Life out of 8 Hours*, 1990, no. 2. "A Break in the Middle of the Show."

A different type of advertisement showed TV sets with other electronic products produced at a factory (see Figure 2.10 in Photo Spread). The ads were plain and straightforward and contained comments about the credibility and history of the modern Chinese factory. A TV set was always at the center of the images; it was presented as the "leader" of the other products. In contrast to the advertisements that showed the real-life role of television in the United States in the 1950s, those in China plainly objectified television. Lynn Spigel's analysis of 1950s TV set advertisements in the United States makes clear that the ads emphasized the role of television in people's lives, by focusing on the "natural" integration of TV sets into daily life. Contrary to the ads in the United States, the ads in China in the 1980s, especially the early 1980s, seldom depicted any people. Although people are shown in the advertisements or other visual portrayals on TV, they are not depicted watching TV. The early ads for TV sets were often drawings rather than photographs. The drawings have clean-cut straight lines and simple colors, which makes the electronic products objects.

The advertisements unified their take on "TV" as a piece of modern electronic equipment while emphasizing in which factory they were produced. The nonspecific representation of TV in these ads was similar to that of an electronic oven or roaster.

In fact, TV sets were in such high demand that people had to get special permission or go through back-door channels to buy one. In the early 1980s, most Chinese people could not buy a TV set even when they had saved enough money to do so. The job of a TV factory manager was more important and busier than that of a mayor because of the high demand for the product. TV factories would negotiate with retailers and ask them to take other electronic products in exchange for the factories filling the retailers' orders for TV sets. Sales of these other products thus affected how much profit retailers could make to sustain and develop their businesses. The advertising of TV sets was not simply for the purpose of selling more TV sets but also was intended to help the electronics industry in general. Many factories could get rid of their unsellable outdated products to activate and sustain their production. The 1980s in China constituted a unique moment when a highly regimented socialist economy was transferring to a newer, partially market-regulated, capitalist economy. During this transition, the relationships between advertising and commodities and between producers and retailers were unusual.

The relationship between consumers and advertising was even odder. The comic (see Figure 2.11) vividly illustrates the disharmony between ideological policy and the real economic situation in Chinese society in the 1980s. In the cartoon, an old man who looks like he is from a rural area asks a sign painter, "Hi, young man, where can I buy this big color TV that you are painting?" The young man answers, "Hi, Mr. I'm only the person who is in charge of painting this ad . . ." The cartoon points out that common people did not know where to buy a TV set, even if they had the money to buy one. However, there was an authoritarian, "official" reason to advertise TV in this manner. In China in the 1980s, TV advertisements served a different purpose than the commercial one we expect advertisements to fulfill; they displayed the productivity and the credibility of the newer or modernized products of the state-owned factory. TV was acting as the savior of the factory and was supposed to draw people's attention not only to itself but also to other electronic products. The biggest and most eye-catching element in advertisements of TV sets from that time is often the bold characters in which the full names of the manufacturer are written. The name is accompanied, of course, by a big photo of a TV-set screen.

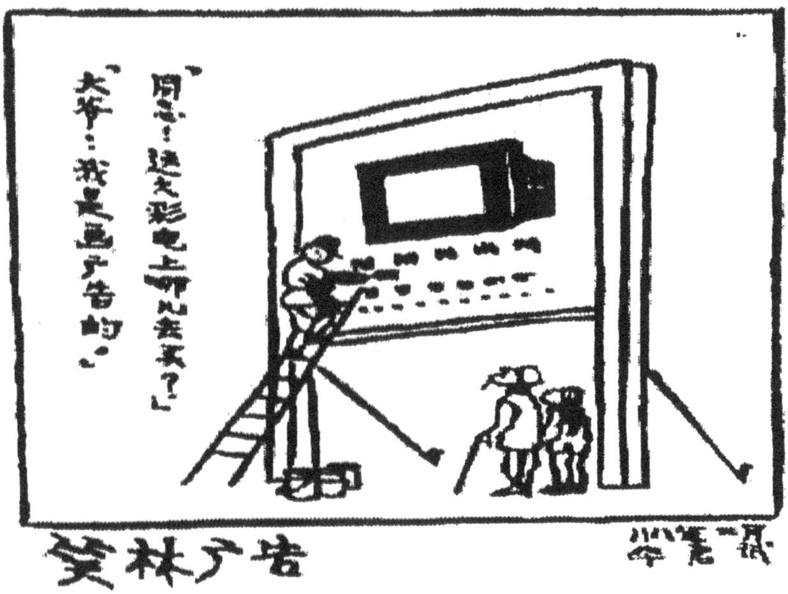

Figure 2.11: *China Daily*, March 29, 1988. "A Funny Advertisement."

These advertisements, which did not provide sufficient information about where to buy the product, thereby emphasized the producer more than the products. TV was produced and promoted as a symbol of modernization more than as a product for common people to actually incorporate into their lives. A TV set was a prestige item not just because of its relatively high price but also because it was a creation of the whole cultural environment. Television was just one element of many that combined to create an idealized lifestyle that became part of the promise of modernization in 1980s China. This promise is very similar to the promise of technology in developed nations like the United States and European nations. However, while the developed countries created models of a modernized lifestyle according to their own "technics," a country like China, driven by the needs of improving the quality of life in its own economic, political, and cultural milieu, could not approach these models for economic reasons. Accordingly, the media were struggling between the two symbolic meanings of having a TV set for the average person: a better life or a higher social status.

Educated by Ourselves and Entertained by Otherness

Raymond Williams (2005) has argued that channel capacity always precedes content; Chinese television stations did not have enough shows to air in the early 1980s. Some factories even offered TV sets as prizes for departments that could produce new TV shows. But the television stations were like a wok on the fire that could not wait for the rice. They had to look for other resources to fill their broadcasting time slots. Therefore, China started importing TV shows and dramas from abroad. Dramas imported from Japan, Hong Kong, Taiwan, Singapore, the United States, Brazil, England, and other countries suddenly flooded the media landscape. The most influential TV dramas in the mid-1980s were the Japanese TV dramas *Zisan Silang* and *O Xin*. The former was about a soldier who returned from World War II and started rethinking the relationship between people and the government, war and humanity. *O Xin* was about a strong and brave Japanese woman who survived World War II and established her own business through hard work and honesty. Chinese audiences were touched by the bravery and courage of this hero and heroine as they faced cruel social conditions. The viewers felt sympathy for the fate of people who had to cope with the hard situations caused by economic conditions and/or governmental decisions. Many Chinese people now in their sixties remember watching *Zisan Silang* with a group of people in front of a small black and white TV set. Several can still sing the show's theme song. In these viewers' recollection, the show actually was not very good from an artistic perspective but they found compelling the depiction of life in Japan after World War II. The show portrayed similarities between Chinese and Japanese cultures in terms of what the war had brought into the common people's lives as well as in terms of how common people in both countries survived and went on to successfully navigate modernization. It was actually an education in patriotism to Chinese people because the drama depicted the bitterness of Japanese people toward their government after the war. To Chinese people, this show was the most powerful evidence that the war was not right. This differed from what many Chinese people previously had been taught.

At the same time, in *O Xin*, the heroine's background as a peasant's daughter who survived extreme poverty to become a successful business owner taught audiences that modernization could bring new opportunities. It showed that people could grab onto the chance to work hard while being brave, honest, and kind. *O Xin* criticized merchants who wanted to cheat people of their money as well as the women who had to sell their bodies to gain a better life in the new, modern urban milieu. These scenarios helped educate Chinese people during their own transition from a society mainly under governmental control to a society oriented toward the economic market. Although there were cultural differences between China and Japan, the TV drama presented the prospect of an idealized future that would follow upon Westernized modernization.

The memory of these two Japanese dramas is clearest to people now age sixty or older. Most people who grew up in China in the 1980s have a much clearer memory of dramas that were imported from Hong Kong, such as *Huo Yuanjia, Zai Shang Hushan Xing, Shanghai Tan, She Diao Yingxiong Zhuan, Yangjia Jiang*, and *Shisan Mei*. These shows contain similar themes of patriotism, masculinity, personal loyalty, and other ideologies typical of the patriarchal culture of pre-modern China. These fictional historical dramas were produced in the late 1970s and early 1980s in a cultural context of narcissism in Hong Kong, based on its speedy modernization process and ethnic consciousness during the late phase of its colonization. These Hong Kong dramas did not endanger the dominant ideology then being promoted by the Chinese government. Instead they worked as an education in the spirit of unification by showing traditional dynastical culture that emphasized the unification of nation. The shows reminded viewers of the darkness that enveloped China at the beginning of the twentieth century when many foreign countries were invading. The selective importation of these Hong Kong TV dramas was obviously under the control of the Chinese government. There was nothing in the media that might challenge the dominant Chinese governmental ideology during the period of early economic reform. Nevertheless, for the first time, Chinese people had a chance to watch dramas that revised historical facts so as to produce more entertaining stories. The heroes and heroines wore fancy and beautiful clothing, had free and dramatic lives, and experienced legendary moments and romantic love. These dramas about early twentieth-century Shanghai actually displayed a highly Westernized and modernized lifestyle, which included churches, telephones, cars, carnivals, parties, and night clubs. All of these things were actually a combination of elements from 1970s Hong Kong and the screenwriters' vision of life in early-twentieth-century Shanghai. Chinese people had never before been able to watch scenes of romance and violence that were full of modern fantasy and influenced by Western and gang culture. It was not until many years later, when

The Godfather became available to Chinese audiences, that Chinese viewers realized that not everything in *Shanghai Tan* was as original as they thought. Chinese audiences sensed the patriotism these shows demonstrated in their praise of the heroes of the nations. The shows both idealized and modernized Chinese martial arts, music, costumes, and even natural scenery. These fictional historical dramas excited Chinese audiences and allowed them to fantasize about a virtual world created by TV.

Accompanying this fantasy, in the 1980s, TV dramas from the United States displayed a culture that Chinese people could barely imagine existed. The detective TV drama, *Hunter*, is too minor and dated for many Americans to remember now, but this TV show had brought into view many elements of American culture in China in the 1980s. Cars, Coca-Cola, drugs, telephones, red wine, TVs, family houses, boats, vacations, books, religions, fashions, fast food, diners, Western beauties: these images and cultures were very modern and exotic to Chinese audiences. The difference between Hong Kong dramas and American dramas was that the shows produced in Hong Kong were overly dramatic and fictional. Those from the United States seemed more realistic, including many desires that corresponded to real life. What appealed to Chinese audiences was not the invisible social system that had created and sustained all the images that they saw in the TV shows, but the objects that symbolized a modernized lifestyle. In American police dramas, the cars were like toys that could be destroyed randomly and replaced easily. All the things that Chinese people cherished and could not get in the 1980s were a normal part of common American people's daily lives. The Chinese government imported *Hunter* because it conveyed a message about capitalist law, social justice, science, technology, and the darkness of capitalism. The plot of the show was very simple. It told how a policeman and his female colleague caught and investigated criminals to protect the safety of common people. However, the show's themes were too remote for Chinese audiences; what attracted viewers was the wealth of a highly technologized, commercialized, and mobilized developed society.

The imported TV programs entertained Chinese audiences with their exotic and dramatic narratives and images. No matter whether the subject was the life story of a Japanese woman, historical fiction, or an American detective, all provided a channel through which Chinese people could imagine the excitement of a modern, commercialized culture. The gatekeeper of Chinese media, the Department of Chinese Broadcasting and Telecommunication, made sure that the exotic culture would inspire thoughts that were not too distinct from the dominant ideology established by Chinese political and economic needs.

Although the amount of Chinese shows was insufficient to satisfy the needs of Chinese TV stations and fill broadcasting time slots, China did produce some

TV dramas, such as *The New Star* (Xin Xing), *A Mouthful of Vegetable Pancake* (Yikou Caibingzi), *The Snow City* (Xue Cheng), and *Idle the Time Away* (Cuotuo Suiyue). However, the goal of these domestic TV shows was always education; the narratives and characters always were presented very seriously. Ideological education was predominant in Chinese-produced TV shows.

Therefore, TV shows in China, apart from those presenting propaganda news and governmental policies and documentaries (which we will discuss later in the next chapter), did not have content that directly related to people's real lives, and only the imported TV dramas provided fantasies that removed viewers from their own lives. Viewers obtained a general idea about modern lifestyles from the American TV dramas and Japanese family dramas. The living standard was much too modernized and characters were too mobile in these imported TV shows compared to the social condition of China in the 1980s, especially in rural China, to be directly relevant. Generally, Chinese viewers were educated by the domestic TV programs and entertained by the imported ones. Imported TV shows were only loosely censored on many municipal and city TV channels. This was justified by the idea that those TV shows should not and could not be taken seriously; they were just entertainment. In fact, both the domestic and imported TV programs were creating an idealized modernized future for Chinese audiences by criticizing the past, creating an idealized present, and promising a better future, a future that would adopt the good aspects of modern Western society while discarding the bad.

TV Village /TV Community

As mentioned earlier, the purpose of having a TV for individuals in the society was part of the general pursuit of the "Four Modernizations." In the early 1980s, the mass media and journals often mentioned the number of families that owned TV sets and then pointed out how many families in a village or in a work unit residential community had a TV set. The number of people owning a TV became an important index of the economic situation, and therefore, of the degree of modernization in the village or the work unit. As argued above, this criterion is not necessarily true; using TV set ownership as a measure of modernization could convey an inaccurate impression of the real economic situation in a particular area. The phrase "TV village" (*Dian Shi Cun*) was applied to villages that, despite high levels of TV ownership, were not necessarily rich.

The concept of a TV village reflects one of the ideological understandings of private and public spaces in the process of modernization. In the United States, from the beginning of the promotion of TV sets, TV had always been portrayed as an item that could unify the family and contribute to domestic entertainment. Television was popular in the United States, Lynn Spigel suggests, mostly because it contributed to and reinforced the unity of families. From a married woman's perspective, a TV set was valuable because it could entice her husband and children to stay at home and also because it provided entertainment while the woman was engaged in tedious housework. In contrast, in China, television's greater value lay in connecting a "small" family, the nuclear family, to the "big" family, the nation, and therefore, to the world. In other words, in China, the TV set was portrayed as an item that could make people more open and encourage them to think beyond the needs of their immediate families. Although in the United States, television also played an important role in connecting the individual family to a wider world and in creating public interest in social issues, the main theme in American advertisements for TV sets was promotion of the domestic life of the nuclear family. This is the biggest difference in the construction of the role of TV in the United States and in China.

The role TV played as a servant/messenger kept the Chinese audiences always informed with the national news information and propaganda. This role of TV decided that having "TV villages" and TV communities was meaningful in many aspects. First of all, the ownership of TV sets showed the economic accomplishment and development of the village. At the same time, it showed the ability of village or work unit residents to become aware of various important governmental policies and issues. Being part of the television-mediated world was equivalent to being part of the "outside" of a local community's life. Creating a TV village or TV residential building was a big political achievement, and it symbolized the economic progress made by the village or the work unit. News articles about the TV village phenomenon were intended to encourage the national audience concerning the progress that the nation was making under the Economic Reform. TV sets were used as a positive symbol when they were owned by collectives or given by collective funding.

The new terms "TV village" and "TV building" reflected the characteristics of television as both an individual entertainment and a collective activity. When every family in a village or a work unit had a TV set, they could watch in their own homes instead of standing in front of a small TV set having to watch exactly the same program at exactly the same time with their neighbors. The individual family could decide when and what it wanted to watch in its own home. However, people living in the same village and working in the same work unit usually had similar daily schedules and similar interests. Not to mention that it was very easy and necessary for them to talk about the TV programs that they all watched at night in their daily chats. At the same time, there were not many choices among the few channels. Typically, people had access to only a few basic channels: CCTV, the provincial TV station, the municipal TV station (in some big cities), and possibly a local closed-circuit television station. Thus, the creation of TV villages and TV buildings improved the viewing conditions of people but did not really change the watching activity that people were used to having when they were sharing the same TV set. Neither did it diversify the content of what people were watching in different families. The ability to make real individual choices did not become available until there were more channels. People could make an individual selection of their daily schedule due to the different jobs and various lifestyles that they gained while heading to the end of the 1980s. For instance, some people owned their own business and did not have to follow the fixed schedule that most working units set up; some people were richer and could afford more entertainments after the Economic Reform.

Aside from the honor accorded to the heads of villages and work units that received the "TV village" or "TV building" designation, the title did not make a big difference in people's daily lives in the collective. Requiring every family to buy a TV set at that time would not lead to an improved economy or even better viewing experiences, nor would ownership of a TV set by every family be evidence of economic growth or intellectual improvement. People were coerced into buying a TV set before they purchased anything else for the sake of the collective's honor. Then at night they all watched the same program, which they discussed the next day at work.

In sum, the real effect of the creation of "TV villages" and "TV buildings" in the early 1980s was that it was easier for people to keep watching TV as a part of their daily lives and was an early step in promoting individual ownership of property. Encouraging every family to buy its own TV set simultaneously was the first step in the transition from a complete collective economy to an economy that allowed private property.

Conclusion

All in all, owning a TV set could not truly reflect people's degree of wealth or modernization in 1980s China. The symbolism of TV sets in the general cultural environment was not simply based on the fact that a TV set was an expansive item for most families in the 1980s. In fact, what the TV set symbolized was based on complexity of various elements. The presence of TV sets in Chinese society in the 1980s started a process that was not evident immediately. In the early 1980s, the TV set was similar to other items of modern technology in that it made the same promise of a modernized future for the Chinese people. The combination of the economic and cultural elements encouraged Chinese people to own a set and then provided them the information on modernization through the programs they watched. Owning a TV set was not evidence of a better or more modernized life. Instead, television created an avenue by which people could learn or imitate a modernized lifestyle. This imitation happened both consciously and unconsciously. The impact of TV on Chinese society and culture evolved into a complex and fascinating phenomenon that displayed how the electronic medium had been made possible by the general technological and ideological environment, how the medium interacted with human agencies, and how common people's lives were a legitimated part of the modernization that was modeled by Western technology and culture.

CHAPTER 3
DOCUMENTING NATURE: TV AND THE REPRESENTATION OF NATURE IN 1980s CHINA

An intensive examination of modernity and modernization typically includes discussion of the complicated relationship between technology and nature. Technology can hurt, and even destroy nature, as can be seen by current serious environmental problems. However, technology also helps human beings to live a better and more comfortable life, protected from the harsh conditions of nature, such as temperature extremes. When city dwellers yearn for a vacation outside the city near mountains or lakesides, they are often seeking fresh air, plants, and flowing water. It is debatable if modernization takes away nature. As it shows, technology creates things that supplant the loss of nature in cities. Moreover, urbanites can maintain their notion of nature through different channels; for instance, there are parks, trees, and various *natural* things still existing in our cities, and these natural objects constitute nature per se within the cities. From a cultural perspective, we cannot deny the role that media have played in the process of shaping and maintaining urbanites' perception of nature.

In her analysis of water in urban spaces, Maria Kaika (2005) argues that the modern city is like a "space envelope (the home, the city and the nature)" that creates a programmatic vision of nature for its residents. However, while the programmatic vision materializes our conception of nature as a separate entity, it also weaves nature and city together more closely into a socio-spatial continuum. Although Kaika's argument does not attribute the source of this vision to the media, an idea of "programmatic vision of nature" in the modern city leads naturally to thinking about the impact of the technological media on people's conceptions of nature in a modernized world.

In this chapter I analyze *Talking about the Changjiang River*, a well-known Chinese TV documentary produced in the 1980s, and the relationship between television and its visual representation of nature within the cultural context of 1980s China. First, I will examine how the Chinese periodical press represents the relationship between television and nature.

Adorning TV Sets with "Flowers" and "Fruits"

In examining nature and advertisements in modern society, Judith Williamson (1978) makes an argument about the relationship between nature and the natural. She agrees with Levis-Strauss's opinion that the representation of nature in modern culture is generally "cooked." Using the example of an advertisement for orange marmalade, Williamson points out that the products of modern technology are often shown in the process of being cooked. In this advertisement, images of an orange being peeled and cut are placed next to a marmalade jar. The product "has appropriated the raw" material and "has taken over" the natural object so that it can stand for the raw (Williamson, 1978, p. 103). This theme is present in many modern advertisements. It is not unusual to see an advertisement in which a cup of coffee made from a jar of instant coffee granules is placed next to a bowl of flowers or an advertisement in which a bottle of body lotion is set against a background comprised of a green mountain with beautiful wild flowers. However, "the product can never quite get back to nature, because it can never be signified as nature, only as natural: 'the natural' is the meaning extracted from nature, and there is an invisible but impenetrable barrier between the two" (1978, p. 122). Williamson clearly explains the difference between "nature" and "natural":

> "The natural" is the meaning given by culture to nature; that it is socially determined and not a fixed quality is shown by the change in what constitutes the "natural" from age to age, throughout history. It becomes the justification for whatever society approves and desires. But precisely because of this reference to Nature as the determinant of what is good, as though it were an independent artier, "the natural" becomes the meaning given to culture, by nature—although it is culture that determines "the natural" anyway.
>
> Thus society works on nature in two ways: with technology, to create manufactured goods, and with ideology, to create symbols of "the natural" which are then juxtaposed with the manufactured goods so that meaning may be exchanged between the two. (p. 123)

Williamson's study illustrates one point in my exploration of the relationship between nature and television in media in 1980s China. Because it is culture that gives a natural meaning to products and because every culture differs, the meaning particular products take on within a culture will help illuminate that culture itself.

As discussed earlier, ads for TV sets in China in the 1980s were different than the ads prominent in China now. The earlier ads can be described as plain or even "dull." They did not even give specific information about where people could buy a TV set. However, even though the ads were simple, the designers and marketers made sure to include plastic flowers and fruits next to their images of TV sets.

In ads for West Lake TV, plastic grapes and leaves hang down, seemingly not attached to anything, and with no specific connection with any other image or information in the same picture. The designer's intention to provide decoration and to balance the layout is obvious. But doing so does not require that extra objects be inserted into the picture nor that any objects that are inserted be grapes and leaves. Why did the designer select plastic grapes as the decoration? Grapes are a common fruit in China. Among the fruits and flowers that can hang from the top of the page, grapes are the most common. Images of grapes are commonly seen on tablecloths and other domestic fabrics in China because the grape symbolizes fertility in traditional Chinese culture. The ability to have more children is considered lucky, according to traditional values. At the same time, grapes symbolize harvest and nature. The image brings some life to the ads even though the grapes are made of plastic. The symbolic meanings of the grapes somehow soften the "dullness" of the ad and, most important, domesticate the television set and serve to compare the visual pleasure that a television set can provide to that which other domestic decorations have provided.

As Judith Williamson (1978) posits, sometimes there is not a direct relationship between the product and the items next to it. However, "images, ideas or feelings . . . become attached to certain products, by being transferred from signs out of other systems (things or people with 'images') to the products, rather than originating in them" (1978, pp. 29-30). In this TV-set advertisement, the symbolic meaning of grapes in traditional Chinese culture is used to suggest a role this new and modern technology might play in the lives of Chinese audiences. Consequently, the television is familiarized in a new cultural context by being paired with familiar codes from Chinese tradition.

Using images of grapes, which were generally used for decorative purposes only, suggests the role television would play. Unlike TV-set ads in the United States, which always emphasize how television exists as part of people's lives, Chinese TV-set ads completely *objectified* the TV set. In the Baihehua (lily) TV ads (see Figure 3.1 in Photo Spread), some plastic flowers were placed near the corner of the TV set and nothing was shown on the TV screen. The plastic flower looks more like a peony than the lily of the brand name. However, the image of the flower, even though we know it is a plastic flower, makes the whole ad more visually appealing.

In Chinese cities, plastic flowers and fruits were popular forms of interior decoration in the 1980s. The placement of plastic instead of real flowers next to the TV sets in these advertisements suggests the consciousness of the visual pleasure that television can provide and at the same time emphasizes, that like the plastic flowers, television is a product of modern technology. The visual pleasure and the image of nature that the people perceive in plastic flowers can also be derived from the TV set. A connection is created between a yet unknown product and a familiar feeling. Just as the plastic flowers are the creation of modern technology, TV sets are also the promise of a modern lifestyle within the domestic space and context. Both plastic plants and television are the mimesis of nature.

At the same time, this type of advertisement suggests that the designers did not think that television would cause big changes in society. They were more concerned about the use of TV sets as decorative objects, and so the designers focused on satisfying the audiences' preexisting preferences. Designers of TV-set ads in China were not like those in the United States in the 1950s, who immediately realized the real role that television would play in society and often portrayed television as a member of the family, emphasizing the interaction between the TV set and the people and their daily lives in the domestic area. Chinese ad designers mostly treated the TV set as merely another decoration until the late 1980s, when there was a change to more humanized advertisements. The belief that the TV set was a decoration probably explains why people often hung paintings, calligraphy, family photos, and various domestic decorations on the wall right behind the TV set. It was often the center of the visual decoration in the domestic area during the 1980s.

During modernization, every culture has to face the loss of "raw" nature and create "cooked" nature. However, the definition of nature is constituted culturally and changes at different historical moments, according to Julia Adeney Thomas (2001). For example, houses, which are built by people, should not belong to nature according to the current meaning of the word. However, a church or a castle that is surrounded by mountains or trees does not seem "unnatural." The man-made buildings seem to be part of nature. Judith Williamson's argument about the relationship between nature and the natural explains the modern person's typical feelings when looking at ancient churches standing among green mountains. This demonstrates that our perception of nature changes over time. Contemporary perceptions of nature are very different from those that people had thousands of years ago. People's imagination of nature has been evolving. Social elites often romanticize nature in their literary narration. Nature is not always beautiful. In fact, nature is not always a friend of human beings; however, even a natural disaster can excite our eyes. There is a dynamic relationship between nature and human beings, who should be able to sense nature through smelling, watching, touching, listening, and tasting. However, since modernizing, people spend more and more time inside and their sense of nature is thus reduced to watching, smelling, and listening to such items as manmade plants, fragrances, paintings, photos, and of course, the television. Human beings have always had the desire to not only conquer nature but to "mobilize" it. In the long-distant past, people painted natural scenery and hung these pictures on the wall or brought back flowers, fruits, or even animal body parts to close the distance between nature and daily life. The flowers and fruits that decorated these pre-modern homes still maintained their natural smell and texture. However, the plastic flowers and other artificial replicas of natural things that people use now have reduced contact with nature to its visual component. This last remaining sense of nature, visualized nature, must suffice to comfort man's thirst for nature maintaining the last link between humans and nature outside of the realm of the imagination/arts. One result is that modern industry has produced many artificial odors to compensate for the lost odors of nature. To maintain the connection between nature and daily life, human beings have always been in the process of exploring the way to "keep" while overcoming the harshness of nature. Modern technology is constantly providing ways to keep the "freshness" of nature through "such adoptions."

Television combines the functions of sound and visual media to connect human beings with nature through hearing and sight. Most important, television provides a lens through which we can watch nature from both macro and micro perspectives, for example, a long shot of mountains and rivers and a close shot of a blooming flower. It can also bring the fluidity of nature into our enveloped modern houses, especially into the box-like apartments of big cities. Television functions like a small window through which urban people can look at the outside. Shows about nature provide content to reinforce and give prominence to the role that television can play in enriching people's lives in an increasingly modernized and industrialized society. Paradoxically, as a product and a promoter of modern technology and commercialization, television also to some degree consoles people for the loss of nature by giving them something to replace it with.

During the process of modernization, Chinese TV advertisements paid attention to the relationship between television and nature; the ideology reflected in the 1980s advertisements exposes the primary understanding of this relationship, based on traditional ideas linking domestic decoration and nature. The close relationship between television and nature is also reflected in the brand names of Chinese TV sets: Lily, West Lake, Golden Star, Panda, Peacock, Peony, etc. Both television and nature are objects of consumption.

While Chinese advertisements for TV sets emphasize their role in providing visual pleasure by evoking images of nature like the plastic flowers and fruits have done in the domestic domain, the presentation of nature on television itself is also noteworthy. While advertisements "mobilize" the basic ideology about the relationship between modernization, nature and technology, television plays a crucial role in connecting modern society back to nature. Before we can examine how television connects modern society and nature, it is necessary to review the history and evolution of the definition of "nature" in Chinese culture.

A Brief History of Nature in Chinese Culture

Raymond Williams (1980) contends, "The idea of nature contains, though often unnoticed, an extraordinary amount of human history" (p. 67). As many scholars do, Williams believes that nature as a concept is historically changing and interrelating with other ideas; however, "the idea of nature makes them [other ideas] seem comparatively simple." Scholars since Williams have considered it vital to understand a culture's idea of nature so as to understand the culture itself. In *Discovering Nature*, Robert P. Weller (2006) examines the different conceptions of nature in China. He argues that the dominant concept of nature in China is derived from the European and American conceptions of nature, which understand it to be the other of culture. Weller points out differences between the idea of nature in modernized China and the idea of nature in traditional Chinese philosophies, such as Daoism and Confucianism, which emphasize the flow of nature in culture and have a cosmic view about nature. By identifying *qi* and *tian*, ancient Chinese philosophers emphasized the importance of creating a harmonious relationship between human beings and nature. Their understanding of nature still exists in traditional Chinese arts, such as calligraphy and Taiji.

Nature is also an important symbol in Chinese literature. From ancient times to the early twentieth century, enjoying and praising nature was often a privilege of well-educated Chinese intellectuals. Nature has been imbued with these intellectuals' melancholy separation from friends and lovers and unfulfilled ambition of serving the government. The intellectuals who wrote detailed descriptions of nature were often those who had been exiled to the distant border cities or were privileged members of the upper class who did not need to worry about their future in the government, enjoying lavish lives because of their family backgrounds in the dynasties.[1] Therefore, descriptions of the Chinese natural landscape reflected either a melancholy and physically distant political experience or a leisure activity belonging only to the well-educated intellectuals and officials. Consequently, "nature" in Chinese history is a concept created by elites, and elites can never be separated from the nation; even the exiles are always well-educated social elites. The literature on nature has always helped people gain a concrete idea about what a nation is. The inseparable relationship between nation and nature is reflected in the term "Jiang Shan" (rivers and mountains), which has always been the symbol of nation in Chinese language.

The understanding of nature in China has changed since the early twentieth century, when Western science and philosophy were introduced. But it was Mao's Marxism that really shook the fundamental Chinese understanding of the relationship between nature, people, and technology. Since the 1950s, the Chinese government's policy to conquer nature was at odds with the traditional value of respecting and cherishing nature. Following Mao's belief that "man's determination can conquer nature," nature was conceived of as an object that could be managed and controlled by human beings. Mao's aggressive methods brutalized the traditional Chinese concept of nature. The traditional understanding of nature as a harmonious relationship between people and the world outside of the body was almost completely changed because of the denial of traditional culture during the Cultural Revolution (1966-1976). Perhaps no other culture has paradoxically both treated nature as an object and also included it as an inseparable part of its culture, as China did from 1950 to 1970. "The Great Leap Forward," initiated in 1958, was a plan dictating a total conquest of nature. The effort resulted in terrible damage to natural resources, landscapes, and the national economy.

Mao's attitude toward nature also had another effect. He collectivized nature, making it utilitarian instead of a privilege reserved for artistic appreciation by intellectuals. Nature was no longer a topic only in the conversations of intellectuals. Rivers and mountains would belong to the people. Therefore, people could express their love of them. Although the philosophical understanding of nature had changed, the symbolic meaning of nature remained very stable in Chinese culture. This philosophical understanding was embedded in the documentary *Talking about the Changjiang River.*

By bringing nature to the TV screen, delivering natural images through the camera lens, and having narrators "teach" audiences to appreciate nature quotes from pre-modern poems and classical Chinese, *Talking about the Changjiang River* and other 1980s TV documentaries made nature accessible to the common person.

Born in the Glorious Decade of TV Documentary in China

"The main content on TV was the documentaries, which had specific themes, well-plotted content, and a clear (educational and political) goal," says Liu Yun (2003) in his examination of Chinese television in the 1980s. He rightly asserts the important role that domestic TV documentaries played in the early stage of Chinese television. Four domestic documentaries had a great impact on Chinese audiences in the 1980s and early 1990s. These were *Sichou zhi Lu* (The Silk Road, CCTV and NHK, 1979), *Huashuo Changjiang* (Talking about the Changjiang River, CCTV and NHK, 1983), *Huashuo Yunhe* (Talking about the Grand Canal, CCTV, 1984), and *Wang Changcheng* (Looking at The Great Wall, CCTV and NHK, 1991).

Talking about the Changjiang River was the most popular television documentary in China during the 1980s and is the most remembered now. But it was not the first Chinese television documentary; it was highly influenced by *The Silk Road*, which was filmed in 1979 and coproduced by Chinese Central TV and Japanese NHK (the Japanese Broadcasting Corporation). *The Silk Road* was a successful media product because of the beautiful images and the detailed introduction of history. The documentary incited popular interest in the Silk Road among Japanese viewers when it was shown in 1980. However, Chinese audiences did not show any special interest in this TV documentary (Guo, 1997). Learning from this experience, Chinese Central TV and Japanese NHK worked together again to produce *Talking about the Changjiang River,* which was aired in 1983. The success of *The Silk Road* in Japan communicated to Chinese documentary makers that foreigners wanted to see and know about China, and inspired them to do more in their documentaries than deliver educational content or propaganda (Fang, 2008, New Beijing).

The Silk Road references several historical records. The film combined current footage with stills in a way that modern Western audiences might associate with Ken Burns. Images of ancient paintings, statues, books, and other historical relics are shown between the shooting of natural landscapes. The first episode of this TV series served almost exclusively as an introduction to Chinese history, focusing specifically on the history of Xi'an. Against the soundtrack of Japanese composer Kitaro's electronic musical piece "The Silk Road," the TV documentary starts with a melancholy and nostalgic emotion, which runs through all twelve installments. Other parts of the soundtrack work to endow the history with a mysterious feeling.

In its depiction of locations of historical significance along the Silk Road, the documentary emphasized the relationship between nature and human culture. Natural scenery was shown by spots of historical significance. Rarely was an object or a scene shown without the story of the humans behind the objects. In one of the episodes, a train is shown blowing dark smoke as it moves across the desert, which is empty except for the slim set of railroad tracks running across the barren yellow field. The desert was portrayed as a natural space of unbearable loneliness. By comparing the mysterious and rich history of the Silk Road with the desolate vision of the landscape along the road as it currently stands, nature was depicted as both a creative force of history and an ancient remnant of that history.

The music and the rich historical narrative in *The Silk Road* express a strong sense of nostalgia. This nostalgia and the later "Sichou Zhi Lu Re" (The fever of The Silk Road) in Japan (Guo, 1997, p. 42) illustrate Marshall Berman's claim about the relationship between history and modernity:

> At a moment when modern society seemed to lose the capacity to create a brave new future, modernism was under intense pressure to discover new sources of life through imaginative encounters with the past. (1988, p. 332)

The success of *The Silk Road* in Japan was not accidental. It reflects that Japan, as one of the leading economies in the world, had arrived at a new stage of modernity, as did other Western nations in the late 1970s and early 1980s.

However, this melancholy feeling in Japan about history, born of the lack of confidence in "a brave future," was at odds with the pursuit of a flourishing modernized nation in 1980s China. The late 1970s and the early 1980s was the beginning of China's look to "a brave future" after eschewing governmental policies of the 1960s and 1970s. Therefore, the lack of response to *The Silk Road* in China was not surprising. The emotions the film expressed about history, nature, and modernity showed the two co-producers' views divergent from each other.

As a result of *The Silk Road*'s lack of popularity in China, *Talking about the Changjiang River* had to express a theme more generally acceptable within China's cultural atmosphere. *Talking about the Changjiang River* includes rich information about history, culture, and nature as well as the relationship among the three and many other phenomena related to modernization, such as technology. *Talking about the Changjiang River* was the longest TV documentary (twenty-five installments of thirty minutes each) in 1980s China. What became known as "The Fever of the Changjiang River" (Guo, 1997, p. 42) incited by this documentary among Chinese TV audiences illustrated its importance in reflecting the current issues of the general population. Thus, an examination of *Talking about the Changjiang River* is crucial to understanding the ideal of modernization as ascribed to by both the dominant ideology and the thoughts of typical Chinese television audiences in the 1980s.

Nature as Nature with People

There are two things about *Talking about the Changjiang River* (hereinafter *The Changjiang River*) that many Chinese people still remember clearly: the theme song, *The Song of The Changjiang River*; and the voices of the narrators, Chen Duo and Hongyun. *The Changjiang River* describes the beautiful and magnificent natural sceneries along the river, starting at its origin. A long shot at the beginning of the show displays a powerfully flowing river that is exciting to the eyes of contemporary audiences. Chen Duo, the male narrator of *The Changjiang River*, says, "When I was doing the narration of the documentary, I thought to myself, people don't have time to visit the Changjiang River, but I can and I'm visiting here for them" (Wang, 2006). Chen Duo was correct that many Chinese people would never have a chance to visit the river, which was introduced in the show as the longest river in China and the third longest river in the world. The audience could, however, see it on television and experience it with the help of the narration.

The informative cultural history of the Changjiang River provided by the documentary personified and politicized the natural scenery. "The Changjiang River has lived in the world for thousands of years," the narrator says at the very beginning of the series. Using allegory, this narration immediately added humanity to the river by revealing the real content of the documentary to be not just about rivers in a natural sense but also about the river as part of the history of Chinese people and culture. *The Changjiang River* emphasized and praised construction and water engineering that Chinese people had done during different historical eras. But unlike *The Silk Road*, which focused on ancient technologies, *The Changjiang River* stressed then-current technological developments. The message seemed to be that history is not just the past but also lives on in the present, which contradicts the notion in *The Silk Road* that history made nature a remnant. What Chinese audiences saw in *The Changjiang River* was a nature that had been conquered, and then culturally reconstructed, and that had become the resource of human beings' productions, arts, and lives. The Changjiang River is a part of nature; however, it is beautiful and mysterious because of the literature written by people. The river has human admirers who can conquer and take advantage of it.

The Changjiang River starts with a long shot of the beautiful flowing river and the green cliffs along its banks. Only the river and cliffs were shown during the first few minutes. After the first few long shots, human beings, historical spots, houses, and ancient and modern technological objects appear and become the focus until the end of the show. In subsequent installments, *The Changjiang River* considered one historical spot after another, following the course of the river. Many things, from the historical background to the geographic situation of the different places to cultural life, including religion, daily life, agricultural production, and technology, are presented as if to seem born by the river. Nature gives birth to the culture and then interacts with the culture. *The Changjiang River* traced the history of developments in technology at different historical moments. The mill wheel, the loom, and silkworm breeding all depended on the river sustaining a harmonious relationship with the natural environment. Modern technology, in the form of the dam, the mechanical loom, the ship, the train, the tram rail, and cable car, liberated people from tedious and heavy work. The show presented modern technology as an improvement of older technology and a further liberation from labor. By praising the technological advances made in both ancient and contemporary China, Chinese TV documentaries reinforced the idea embedded in traditional Chinese water paintings: "unlike a Western tradition that tried to capture a pure nature apart from human beings, people form an inevitable part of the Chinese landscape" (Weller, 2006, p. 22).

Technology as "Bridge"

The Changjiang River suggests that nature is imbued with culture. Nature and culture cannot be separated; rather they depend on each other, because nature sets up the context for the birth and development of culture, and culture enriches nature. The predominant theme of *The Changjiang River* is the unification between nature and culture, with technology as a crucial factor in building up a "friendly" relationship between the two and even embedding culture into nature. This melding of nature and culture echoes Raymond Williams's argument (1980) that "in this actual world there is then not much point in counter posing or restating the great abstractions of Man and Nature. We have mixed our labor with the earth, our forces with its forces too deeply to be able to draw back and separate either out" (p. 83). While Williams's argument reflects the worries of a modernist, *The Changjiang River* expresses an eagerness to promote the modernization of China. "What the Changjiang River brought to us would never be a disaster, but would forever be warmth and power," the narrator announces at the beginning of the first episode. Modern technology had tamed the river and would help people benefit from the river.

In Kaika's analysis (2005) of the Promethean project, she asserted that water was central to modern large-scale infrastructure projects because "rivers were harnessed from miles away to feed the growing urban population or dammed to provide electrical power for industry, while mountains were pierced to provide transport routes for people and trade" (p. 107). Besides their contribution to capitalist expansion, "these projects heralded a new relationship between human beings and nature, between nature and the city. Instead of being fearful and threatening, nature became tame and serviceable, a prerequisite for development" (ibid.). Kaika's (2005) argument about the Promethean project shares the ideology expressed in *The Changjiang River*. However, in *The Changjiang River*, the tools and technologies used to tame nature and make the river serve people were not represented as part of an effort to re-conceptualize the city "as a realm outside the reach of nature's processes" (ibid.). Instead, city and nature are connected as the child and the mother. None of China's prosperous cities, among which Shanghai is the best example, would have existed if they had not been located next to the river. The prosperity of Shanghai, as the narrator said, "wouldn't come into being without the Huangpu Jiang (the last branch of The not represented as part of an effort to re-conceptualize the city "as a realm outside the reach of nature's processes" (ibid.). Instead, city and nature are connected as the child and the mother. None of China's prosperous cities, among which Shanghai is the best example, would have existed if they had not been located next to the river. The prosperity of Shanghai, as the narrator said, "wouldn't come into being without the Huangpu Jiang (the last branch of The Changjiang River)."[2] The documentary did not present cities as creations separate from nature but as creations made possible by nature and intertwined in a harmonious relationship with nature.

The documentary's presentation of the coexistence between older technology and newer technology along the Changjiang River makes everything, including the technologies, seem "natural." There is no outside force exerted on nature, pressuring it to change. People could enjoy their lives with the help of technologies that were invented along the river. Kaika (2005) says that the Marathon Dam "became a symbol of a new era in the relation between society and nature, whereby nature ceased to be a source of fear and anxiety and became instead the prerequisite for modernization and progress" (p. 130). While *The Changjiang River* suggests there is a need for technology, such as the famous Gezhouba Dam, to quell the fear and anxiety caused by nature's power, the documentary also describes nature's breathtaking beauty. The spectacular sight of the water passing through the dam is praised as the creation of the modern water conservancy project. The dam is portrayed as a great creation of human technology. The narration suggests that it is strong, durable, and productive. These characteristics of the Gezhouba Dam are also the characteristics ascribed to the Changjiang River throughout the series. The narration about the Gezhouba Dam suggested that great technology not only tamed the wildness of the river, making it beneficial for people, but also preserved the original character of the natural landscape along the Changjiang River. This paradoxical articulation of modern technology and the changes it made on the natural landscape legitimized the relationship between technology, nature, people, and culture within the modernizing context. Nature, technology, and human beings are connected together smoothly and "naturally," the documentary argues. They create a history and enriched culture that cannot be separated from each other.

Therefore, when viewers of the documentary saw the coexistence of traditional lifestyles in remote areas where people were enjoying their lives and in the modernizing cities where technologies were helping people obtain easier lives, the message conveyed was one of diversity, not the uniformity that modernization was supposed to carry with it. The traditional culture was not "ruined" in the process of modernization; instead, modern transportation and communication brought this culture to other places in the world and therefore motivated the

Technology has bridged the distance between modernized big cities and distant villages. Both the technology described in the TV documentary and the documentary itself contribute to a national integration.

Nature and Tourism

The Changjiang River brought into audiences' homes two main types of places: "distant, peaceful" areas and "undeveloped" areas. They included places such as Jiuzhai Gou, in southwestern China, and developing cities, such as Chongqing and Shanghai. Most of the cities, towns, and villages mentioned in the documentary are on the government's 1982 list of spots recommended for tourists.[3] *The Changjiang River*'s detailed introduction to the history and culture of different locations along the river emphasizes developments that took advantage of river power, promoting the idea that the Changjiang River is a resource for the economic boom; the distant and undeveloped places along the river could be a resource for tourism, and the developing cities could further grow through the exploration of new technology, energy, or economic projects. The documentary's confident message that a clear line can be drawn between places of historical significance and places of geographical significance displays the general attitude concerning development in the 1980s. It legitimized the role of nature in constructing people's lives and maintaining the geographic and cultural characteristics of certain areas in China. The documentary is a celebration of the possibilities that technology provides to people. It reinforces the idea that it is best to respect the unique character of the area. Technology could conquer nature while nature could make this conquest unnecessary and worthless.

The documentary expresses an objective, scientific, practical view of the management of natural resources. It also lays out an agenda for the future of various areas in China. The famous coastal cities, among which Shanghai and Chongqing were two important examples, along with other cities already in the process of modernizing were on the fast track to an even more modernized future. On the other hand, the narration in the documentary emphasizes the necessity of protecting the natural environment of China's remote places, such as Jiuzhai Gou National Park in southwestern China. Its magical natural scenery is portrayed as the necessary *alternative* to modernization in 1980s China. The documentary stressed that modernization processes in China were focused on working with the natural geographic characteristics and the cultural history of different places while not attempting to force change on any area. The places shown in the documentary have continued to follow the ideology embedded in the documentary: a modernization made possible by nature. When people today discuss environmental problems and serious pollution in Chinese cities, they tend not to acknowledge that these problems were already an intellectual concern in 1980s China. The government or people did not ignore the potential environmental damage, but in order to develop economy and strengthen the nation, some areas have to sacrifice their natural environment. The development in China was unbalanced because of the characteristics of different areas in the 1980s. A recent study, *Tourism and Modernity in China,* written by Tim Oakes (1998), provides a new insight into the different methods of modernization in different areas of China, such as the effort in southwestern China to preserve a primitive lifestyle and culture to attract tourists. In the following section, I will examine in more detail the paradoxical applications of modernization reflected in *The Changjiang River*.

The Potential Goal of the Documentary

Documentaries are supposed to reflect reality. But "reality" is a problematic term. What is represented in a documentary is always an interpretation of the real. As Richard Kilborn and John Izod (1997) have explained:

> Documentaries can never be any more than a representation or an interpretation of events and issues in the real world. In other words, for all their claims to present the world as it is and their attempts to engage the attention of their audiences by the force of their argument, documentaries can never attain the level of objectivity to which they sometimes aspire. (p. 5)

So, which reality does *The Changjiang River* want to represent underneath all of the images on the screen? From the viewer's perspective, "there will usually be a link between the world that viewers know from their own life experiences and the world that is revealed on screen" (Kilborn and Izod, 1997, p. 39). In other words, even if the documentary is supposed to represent reality, it has to be made with a view to the inherent knowledge of both the producers and the viewers; this is necessary in the encoding and decoding process, as Stuart Hall famously argues. This characteristic of the documentary form requires a more detailed examination of the social context in which *The Changjiang River* was made.

Guo (1997) has pointed out that, "*Talking about the Changjiang River* 'composed' a beautiful patriotic carol in the upsurge of the popular music from abroad." As mentioned earlier, mountains and rivers have always been symbols of the Chinese nation and its states. Thus, it is not surprising that *The Changjiang River* would express patriotism. "The Changjiang River, the Great Wall, the Huang Mountain and the Yellow River, they are heavier than one thousand kilogram in my heart," sang Zhang Mingmin, a Hong Kong performer, in "My Chinese Heart." Actually, it might be very hard for people to talk about rivers and mountains without a patriotic sensitivity because this symbolic meaning has been ingrained in Chinese thought so deeply. The interesting question is how the documentary attempts to engage the attention of its audiences by the force of its argument.

Evernden (1992) argued that to return to things themselves is to observe them before they were "nature," that is, before they were captured and explained, in which transaction they ceased to be themselves and became instead functionaries in the world of social discourse. Once named and explained, they become social creations and their primordial givenness is subordinated to their social utility. Evernden's argument is convincing if we look at the alternative definitions given to nature in different cultures in different eras. Despite nature's complexity, media have tried very hard to create a common consciousness about what nature really is. In *The Changjiang River*, the producer did not attempt to give nature an independent identity. If Evernden's intent is to remind people that nature is a social creation, the Chinese documentary does not seem to think such a reminder is necessary. For thousands of years, nature has been a part of Chinese intellectual life. *The Changjiang River* based its view of nature on this traditional social construction of nature; it was a celebration of nature, in part, as human creation. Consequently, the TV documentary engages the attention of the audience through its choice of objects to shoot and its decision to portray "facts" that are based on well-accepted values in 1980s Chinese culture.

The four most influential television documentaries in China in the 1980s focused on four major landmarks: the Silk Road, the Grand Canal, the Great Wall, and the Changjiang River. One might wonder why the Yellow River was not chosen instead of the Changjiang. They are both symbols of Chinese civilization. They are both often metaphorically cast as the mother of the Chinese nation. In fact, the Yellow River is the real origin of ancient Chinese civilization and for that reason it is also called "the cradle of Chinese civilization." It is very difficult to know exactly why the documentary filmmakers made the selection they did. However, a comparison of these two biggest rivers in China might give some insight into this selection.

The Yellow River flows through nine provinces of China; the Changjiang River also flows through these, as well as through Shanghai, which is at the end of the river. The Changjiang River is almost 1000 kilometers longer than the Yellow River. It would seem too arbitrary to highlight the Changjiang instead of the Yellow River just because the former is longer. There are, however, other significant geographical differences between the two rivers. The Yellow River mainly flows through the provinces of northern China: Qinghai, Sichuan, Gansu, Ningxia, Inner Mongolia, Shaanxi, Shanxi, Henan, and Shandong. Although the cultures of these nine provinces vary, they all belong to northern Chinese culture. Paradoxically, all these areas have been struggling from water-related problems as their populations have been growing. Flooding often occurs, which is disastrous for agriculture. The Yellow River civilization is delicate and has been widely studied by many international scholars, such as Karl A. Wittfogel. It is likely that northern China has been a major topic of interest to scholars because almost all the main dynasties built their capitals in northern China. Most of these cities were located near Shanxi, Henan, and Beijing. Scholars, domestic and international, have learned much about the culture of northern China, especially since the defeat of the Qing Dynasty (1616-1911 A.D.) in the early twentieth century, when many foreigners went to China. Therefore, to create a documentary on the Yellow River would be to rehash an already well-known topic.

The history of the Yellow River is about the transitions between dynasties, which was no longer a major concern of China in the early 1980s, when modernization became the primary national goal. Chinese people in the 1980s knew much less about the diverse cultures and natural scenery of the nine provinces across which the Changjiang River flows than they knew about northern China. Mountains, which blocked communication between the south and north, at the same time provided relatively conscribed and independent environments in which different ethnic groups could develop their own cultures. In the 1980s, there were many places along the Changjiang River, such as Jiuzhai Gou in the Sichuan province, in which people were still living a very pre-modern lifestyle. At that time, in the western Hemisphere, people were already tired of the pressures and burdens of their modernized lifestyles, and were looking for places to which to escape. In southern China at that time, there were still many people unused to electronic products. Modern cities such as Shanghai, undeveloped distance villages such as Jiuzhai Gou, and very religious historical spots such as the Leshan Buddha Statue were all locations to which people of northern-centric Chinese culture had not been exposed. Presenting Changjiang reminded audiences of places along the Changjiang River that ancient intellectuals had described and praised, even though the river had been nearly ignored by dominant Chinese culture since the early twentieth century. The social chaos that China had experienced since the late nineteenth century had drawn people to pay too much attention to northern China, which is the political and military center. When China decided to open the door and welcome foreign investment, places with less political complexity, like southern China, drew new attention. Thus, in the 1980s, talking about the Changjiang River made much more sense than talking about the Yellow River.

Moreover, *The Changjiang River* was a coproduction of Chinese and Japanese media producers. The targeted audiences for *The Changjiang River* were not only Chinese viewers but also Japanese viewers. Talking about the Changjiang River was like opening works of literature and reading the other half of books that had been ignored for such a long period. Most of these places were treated in literature as mere places of exiles for officials and intellectuals. The traditional Chinese literary works often expressed disappointment or anger with the areas, but all the natural sceneries were recorded with a complex blend of melancholy and escapism. Distant from the northern political center, the Changjiang River drainage areas preserved and spawned a diversity of cultures. In other words, to talk about the Changjiang River is to represent a China in possession of diverse cultures and alternative resources that could be employed for a modernized future.

The second aspect that can help us understand the reason for producing the documentary is that *The Changjiang River* is one of four documentaries shot by Chinese Central TV in the 1980s. The choice of the four cultural marks—the Changjiang River, the Great Wall, the Great Canal, and the Silk Road—all share the symbolic meaning that Kafka claimed when he talked about the Great Wall. According to Kafka (1971), the Great Wall was built not for the purpose of driving away the northern invaders but as a symbol that was intended to strongly construct the feeling of belonging in the people living south of the Wall. Kafka's argument about the role of the Great Wall may hurt the feelings of most Chinese people, but his view makes us think more about the rhetorical role that the Great Wall has played in Chinese culture. For most Chinese in modern China to say that the Great Wall is a symbol of Chinese history and nationality is very reasonable. Kafka's argument might evoke vehement protests from Chinese people who are really familiar with Chinese history. Whether or not the Great Wall was really built for blocking the invaders or constructing domestic unification within the dynasty cannot be examined here. However, this discussion really stirs up the already complicated explanation of "nation."

The Changjiang River has a symbolic meaning similar to that of the Great Wall, although the Changjiang River is much more "natural" than is the Great Wall. To talk about them in words, according to Evernden's theory, already places them in an even position. Like the Great Wall, the Changjiang River has always been symbolic of China as a nation. The tone of the documentary is inviting. It praises forms of modern transportation by which people can travel to distant places and celebrates the achievements that the bigger cities, such as Shanghai, gained as a result of modern technology and development. Unlike the Great Wall, the Changjiang River does not block anything; on the contrary, it integrates places and cultures of various geographical characteristics. In the 1980s, the symbolic meaning of the Changjiang River is a more appropriate symbol of the Chinese nation than the Great Wall.

The Changjiang River echoes the ideology of *The Silk Road*, but it presents not only the river's past but also current technological uses of the river's power. When discussing the bifocality of modern culture, John D. Peters has argued, "What is uniquely modern is the claim of indexical verifiability embedded in our representations of social totalities . . . Part of what it means to live in a modern society is to depend upon representations of that society. Modern men and women see proximate fragments with their own eyes, and global totalities through the diverse media of social description. Our vision of the social world is bifocal" (Peters, 1997, p. 79). What Peters suggests is that our understanding of society is based on a bifocal reception of the local along with the totality of a bigger concept, such as a society, a nation, or a world. The documentary's selective and detailed introduction to different sites along the Changjiang River displays the locality of each particular place and at the same time, following the flow of the river, gives the audience a neat and complete description of a unity of people who share the same origin and drink the water from the same river. Moreover, this specific unity is facing a bigger other, the world, as is clearly suggested in the documentary's narration about Shanghai at the end of the film's sequential narration. The Changjiang River is a reality; however, for most audiences "the world beyond the local exists as a visible totality only in discourse and image, though its fragmentary and scattered effects are all too evident in the lives of flesh and blood people" (Peters, 1997). Television, as a new medium in Chinese people's lives, increased the visibility of this totality and unity of the society and the nation, and even the world as a whole.

Another interesting question raised by the documentary is related to Karl A. Wittfogel's thesis that no study of Chinese culture can ignore the fact that Chinese have a "peculiar geo-agricultural past" (Wittfogel, 1970). Wittfogel argues that Chinese culture is a "hydraulic culture," in which food production depends on large-scale cooperation, with dominant social and political power remaining in the hands of "the leaders of the commonwealth who direct its vital external and internal activities—military defense and maintenance of peace and order" (Wittfogel, 1957). According to him, this is one of the main factors that made the single-centered unified Chinese government possible and therefore provided the conditions that gave rise to communist China. Wittfogel's "hydraulic-power determinism" deserves more examination. Despite its overly determinist Marxist view, Wittfogel's thesis provides one of the perspectives from which we can look at the uniqueness of Chinese agrarian culture. "Unification" is one of the essential characteristics of communism. *The Changjiang River* conveys the importance of maintaining a unified country, but this idea of unification is different from what was expressed from the 1950s to the 1970s. *The Changjiang River*'s representation of differences among different places declares the ideal of development that respects the uniqueness of various locales. Because the documentary represents the objects and sceneries following the flow of the Changjiang River, it feels like walking along the river and things appear on the screen in a natural order, but we all know that the media can only represent a small part of reality. In other words, media's selective representation of reality is always rooted in an invisible plot. The documentary tries its best to display a river that has bred a diverse culture. It reveals an economically unbalanced China and rationalizes this unevenness as the result of geographic differences. By showing the beauty and comparatively primitive lifestyles of people in the mountains and remote villages, the television documentary makes visible "an invisibly vast society" (Peters, 1989).

To both village residents who happened to have TV sets and urban dwellers who comprised a large percentage of TV-set owners in the 1980s, the television documentary communicated an alternative vision of China. The rivers and mountains as symbols of the nation were unified by the flowing water. The documentary's preoccupation with this theme suggests that the process of unification that Wittfogel speaks of required overcoming inner divisions to develop a modernized and internationalized China. Remote, primitive parts of China were intentionally "protected" from the kind of modernization that Shanghai was experiencing by both the film and government policy. The tourist promotion embedded in the documentary showed the influence of modernization in developed countries. If Wittfogel's argument can still apply to Chinese politics, it must be applied with a complexity that addresses these new conditions. As the documentary reveals, this newness was caused by the negotiation between a unified system and need for diversified economic forms. At the same time, by showing natural scenes along the banks of the river, the documentary emphasizes the necessity and the beauty of maintaining a diverse culture in China. Maintaining the uniqueness and authenticity of areas of minority lands was done not only out of respect but also out of a need to preserve a "natural" and alternative vision amid the trend toward modernization and urbanization. It was easier, as well, to develop these places according to a model different than that by which the big cities had been developed. No other medium could have competed with television in presenting a vivid picture of the economic disunity according to the geographic characteristics in the 1980s.

Conclusion

Combining views of the prospectus, technologically advanced cities, breathtaking natural scenery on top of mountains, and rich cultural artifacts, the television documentary *Talking about the Changjiang River* imagined a vast nation full of diversity and comforted audiences feeling the loss of nature as a result of modernization. By representing a beautiful vision to connect television audiences with nature, *Talking about the Changjiang River* portrayed an idealized harmony between the gains and the pains of modernity. The line between urban areas and remote villages was drawn. The most important point made in the documentary is that the division between urban areas and distant areas was not the result of an unbalanced degree of culture and knowledge but of the natural difference between areas. This is also one of the reasons why rural areas were seldom mentioned in the documentary. The rural areas were between the big cities and the remote areas, and the film's occlusion of these areas reveals an idealization of modernization. Standing between big cities and remote villages, the developing rural areas exemplified the complex relationship in the 1980s between technologized culture and agrarian culture. Weren't these rural areas along the river in the 1980s? By eliding the rural areas, the documentary argued that both participating in and escaping from modernization were options because big metropolitan areas and primitive villages were harmoniously developing at their own speeds. Highly developed, modern, capitalist society seeks to heal the ills of civilization through nature; thus, *Talking about the Changjiang River* showed both the eagerness to modernize and the need to seek a panacea from the ills of modernization through its exploration of "nature." The fear of losing nature and tradition and the eagerness to catch up to highly developed modern nations were harmonized. The complex outcomes of modernization were concealed by the film's representation of the improved living conditions in cities creating peaceful sanctuaries in the mountains. Modernization was romanticized by the film's selective presentation of highly urbanized and remote places along the Changjiang River.

Television documentaries in the 1980s portrayed a China well prepared to modernize. Obviously, China had long been on track to connect, simultaneously in many ways, with the rest of the world in the 1980s. As we all know, modernization had brought serious consequences into developed countries by the 1980s. These consequences, such as the consumer culture and turbocharged materialism, when they were introduced to China, became part of the exotic culture in the 1980s. What would happen after these consequences came into an exotic place through the time channel: from the 1950s and 1960s United States to the 1980s China? *The Changjiang River* was too excited about and busy with the "present" and its conceptualization of the future to consider that issue. Chinese TV audiences' "programmatic vision" of nature is a product of the interaction between television and its cultural context.

Notes

1. Around 220-280 A.D., during the eras of Wei, Jin, Three Kingdoms, and South and North Dynasties, there were many upper-class bureaucratic families whose descendants did not have to work for their luxurious lifestyle. They appreciated philosophies of nature and wrote many literary works about nature.

2. This was in the last episode of *Talking about the Changjiang River*.

3. These are listed in Wen and Tisdell's (2001) detailed examination of tourism and Chinese development.

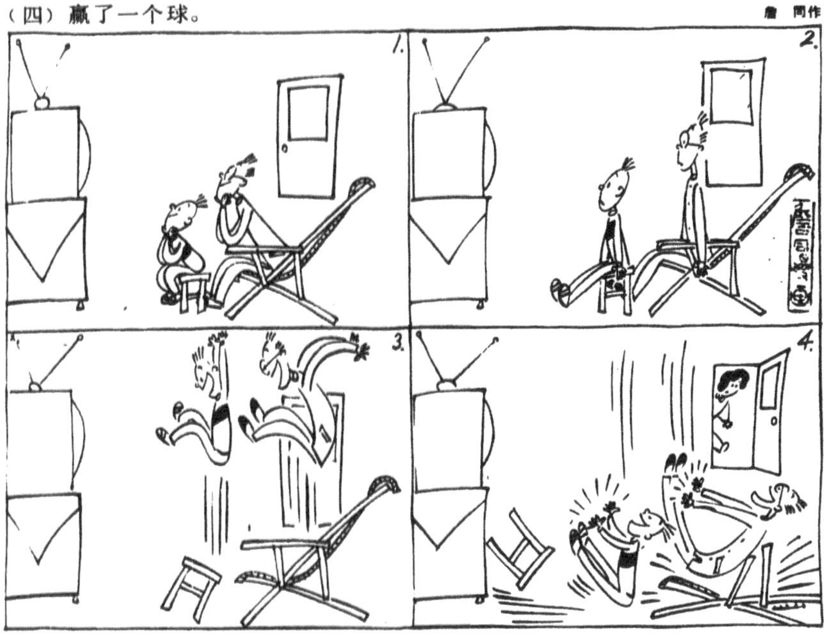

Figure 1.5. Shanghai Pictorial, 1985, no. 6.

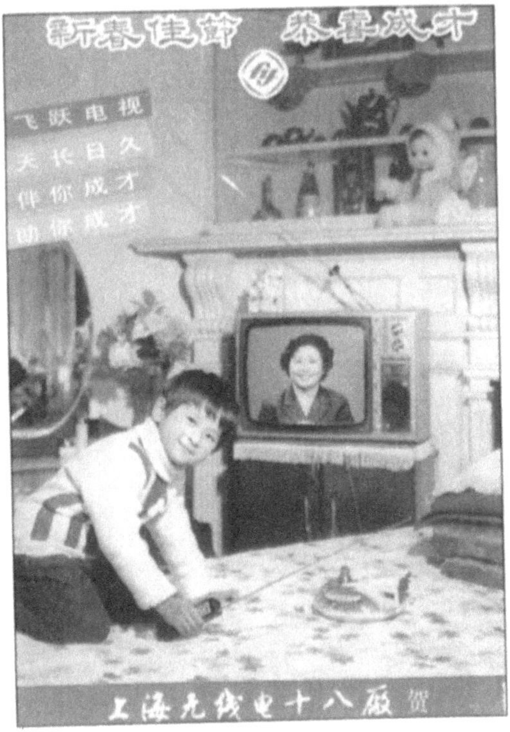

Figure 1.6. Popular TV, 1984, no. 2.

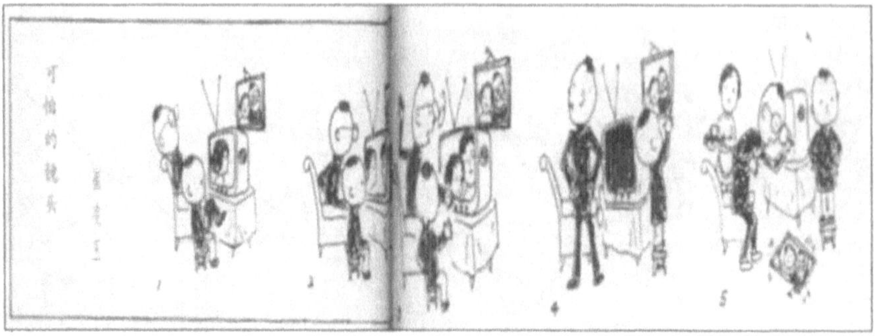

Figure 1.7. Popular Cinema, 1980, no. 5.

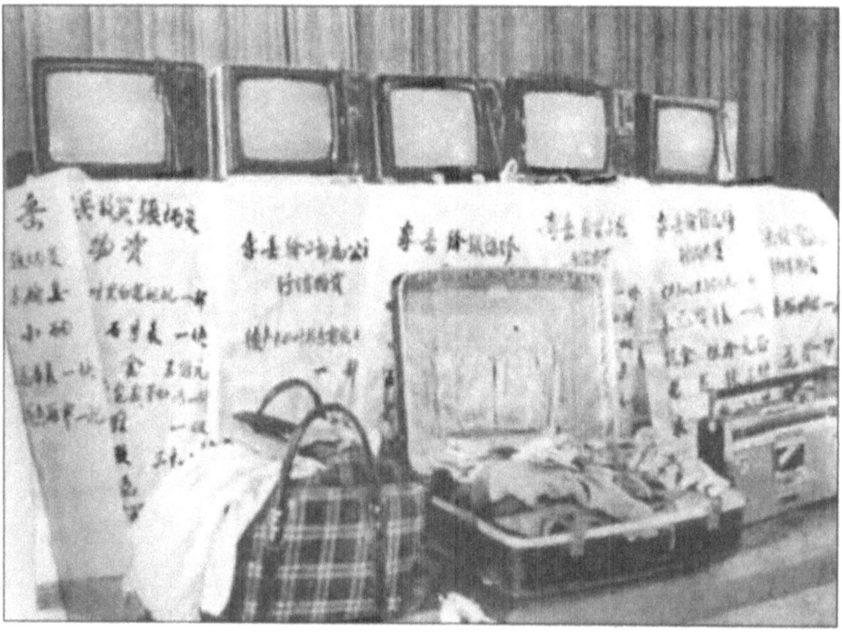

Figure 1.10. This photo shows TV sets being used as bribes to corrupt governmental officials.

Source: Chinese Pictorial, 1981, no. 4.

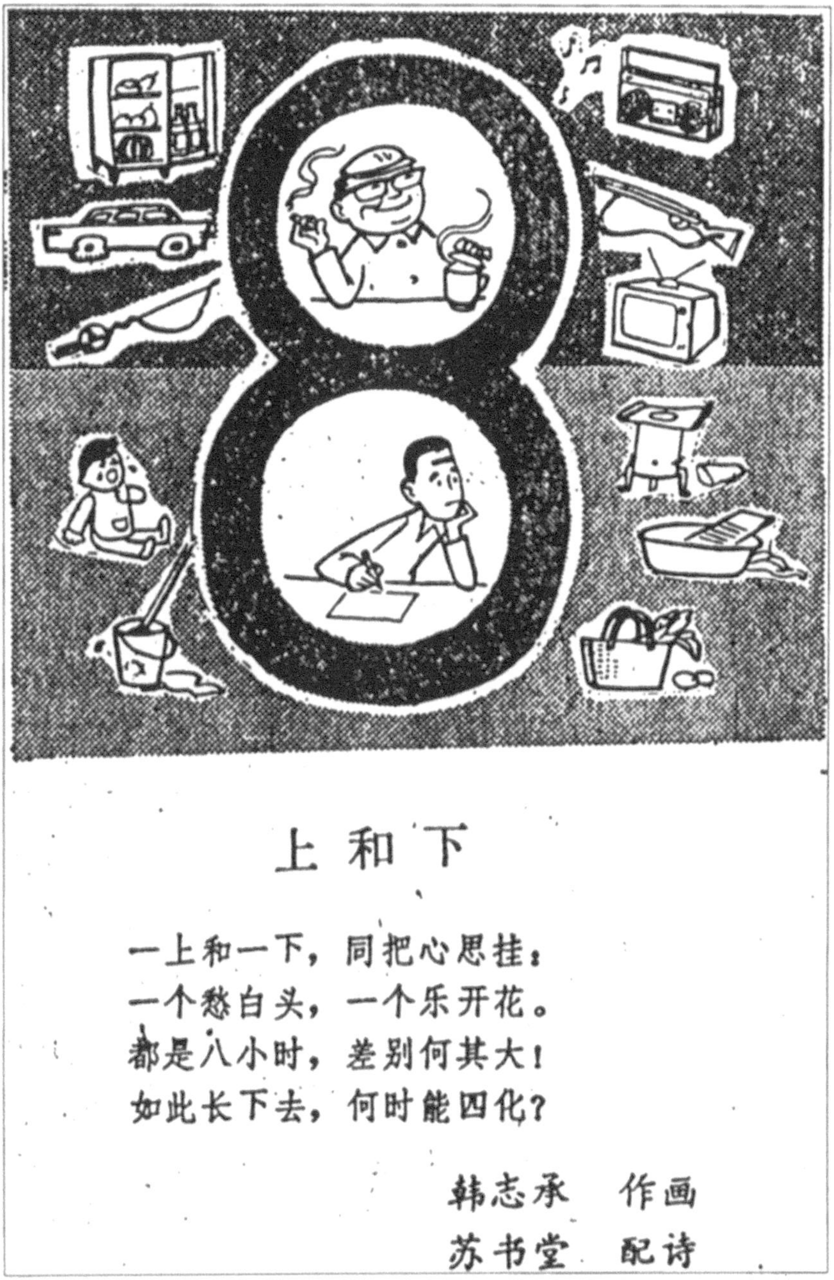

Figure 1.11. In this cartoon, an "8" divides two types of people: a governor who is on the top and enjoys watching TV, hunting, listening to the radio, fishing, and having a refrigerator and a car; and an intellectual who has to take care of his child and household chores. A tea cup in the official's hand and the pencil in the intellectual's create an ironic contrast.

Source: Life out of 8 Hours, 1980, no. 6.

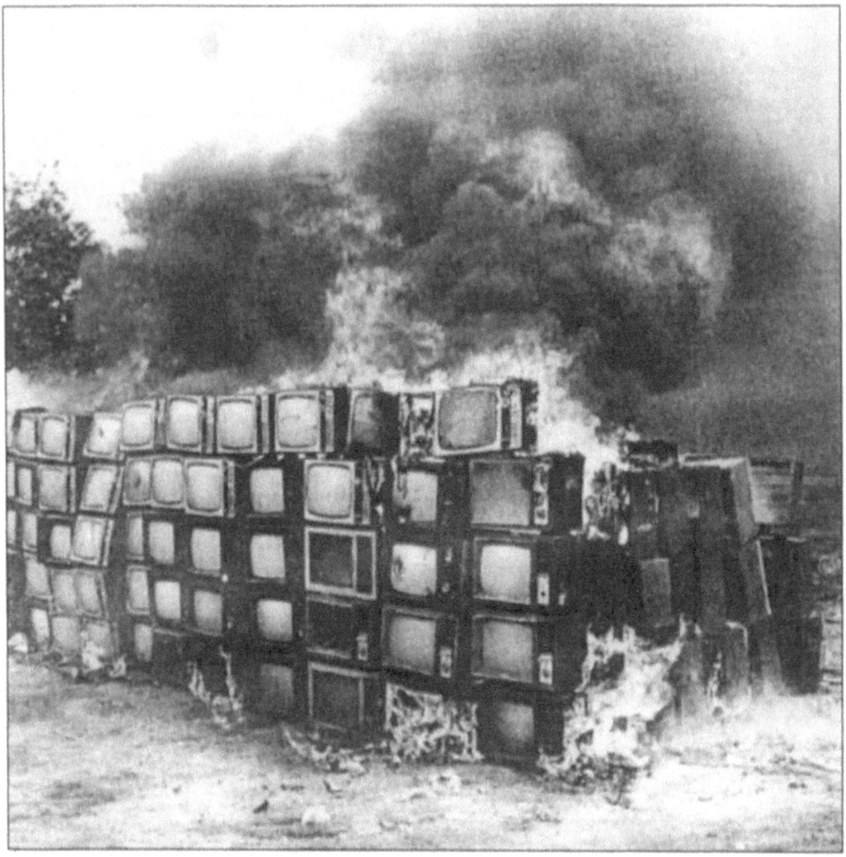

Figure 1.13. This photo shows how the Commodity Inspection Bureau destroyed disqualified TV sets. TV here is completely objectified. TV sets are destroyed in the same way as shoes and bags.
Source: Shanghai Pictorial, 1990, no. 1.

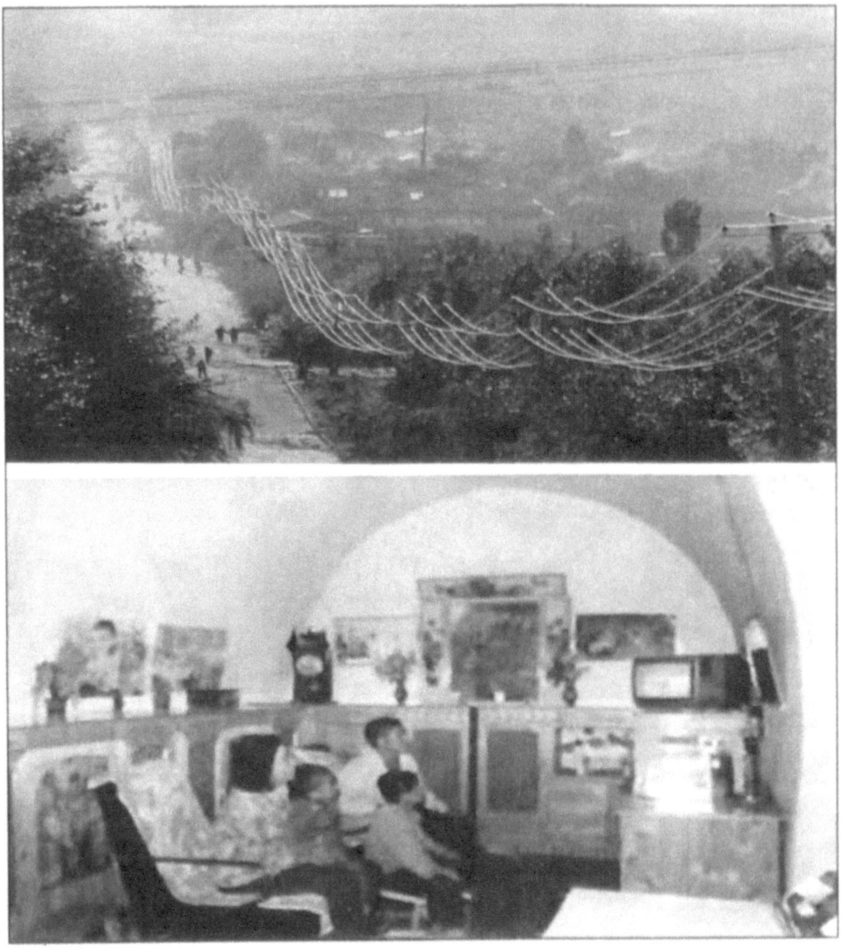

Figure 2.1. In an article introducing the modernized city of Yanbei, pictures displaying electricity construction and a family watching TV are put next to each other.
Source: Chinese Pictorial, 1984, no. 1.

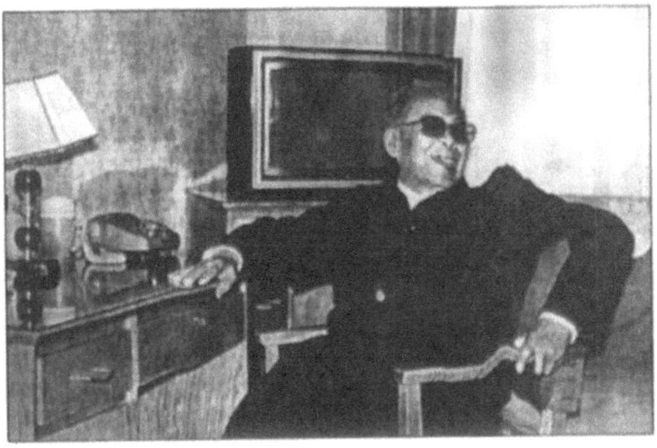

Figure 2.2. Professor Zhou Gucheng, a well-known historian, is shown sitting in front of a TV set during an interview.
Source: Shanghai Pictorial, 1984, no. 1.

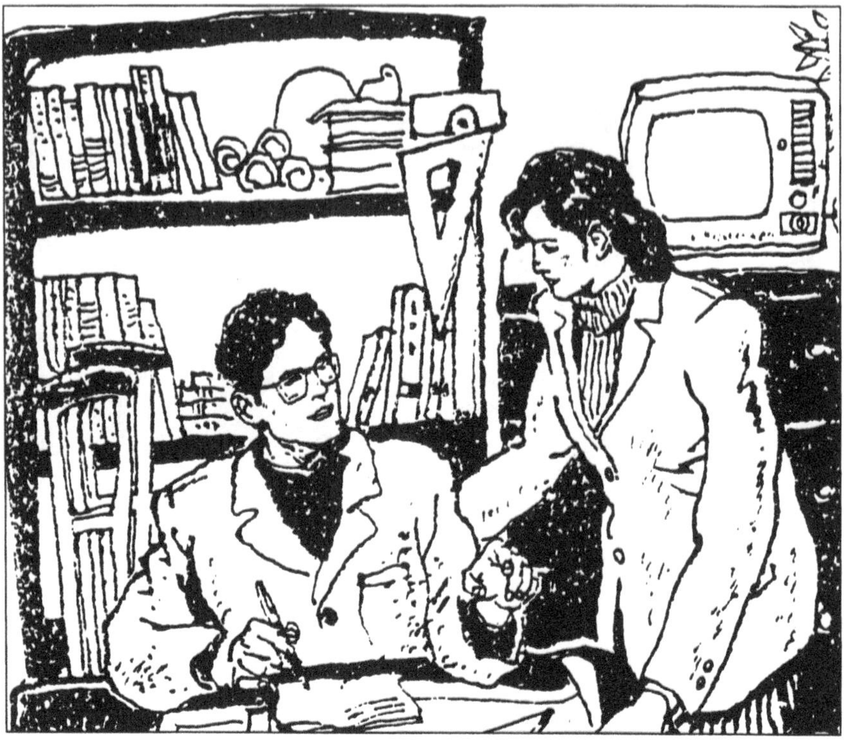

Figure 2.3
Source: Modern Family, 1985, no. 2.

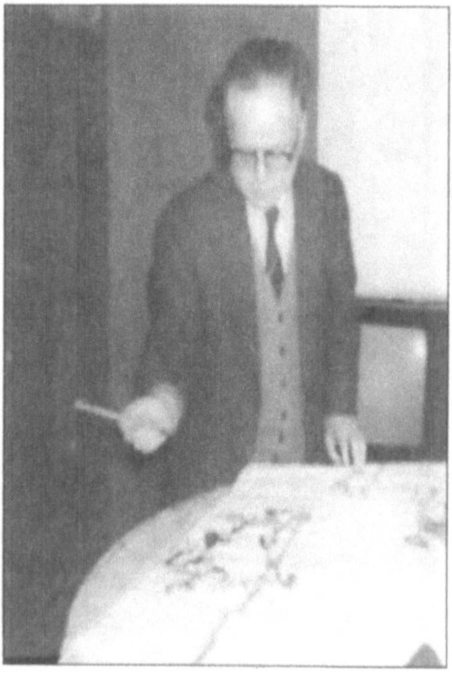

Figure 2.4. An accomplished Chinese painter is working with a TV screen behind him.
Source: Chinese Pictorial, 1984, no. 4.

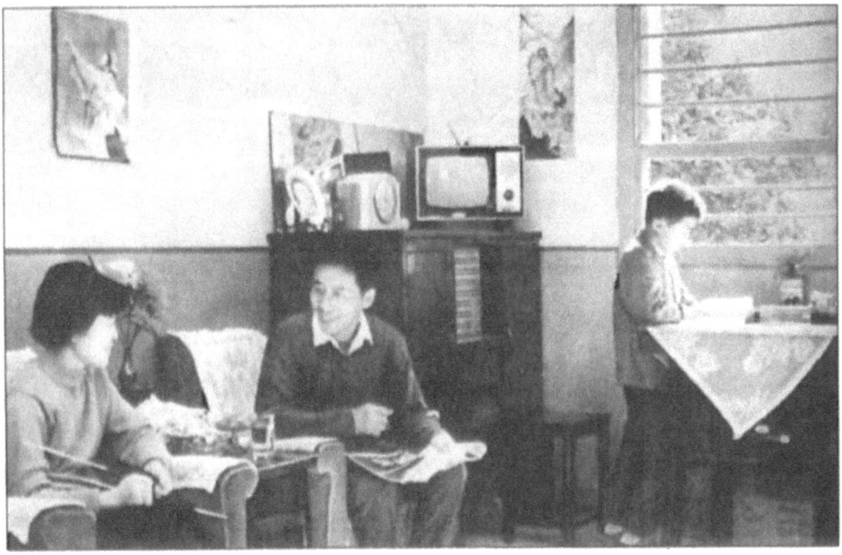

Figure 2.5. This is a photo in an article about the standard of living in a worker's family in the era of the Economic Reform.
Source: Chinese Pictorial, 1981, no. 5.

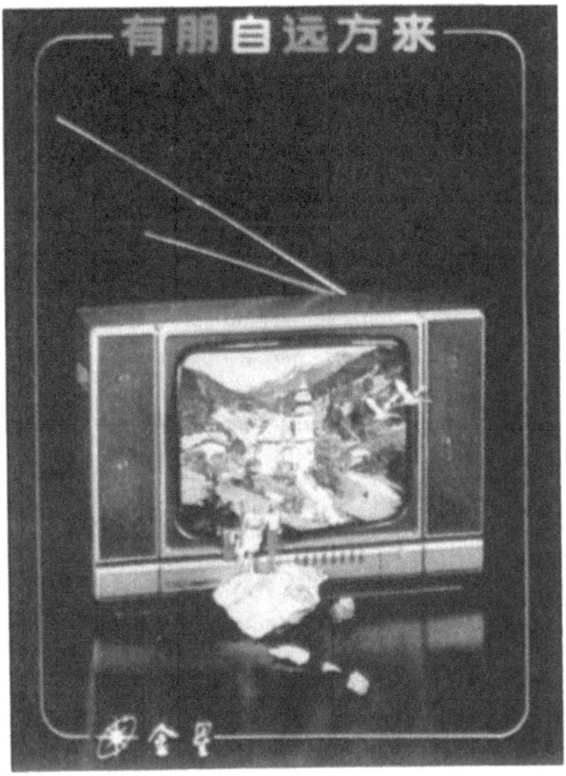

Figure 2.7. "Don't you feel happy when your friends are visiting you from a faraway place?"
Source: Chinese Advertisements, 1986, no. 2.

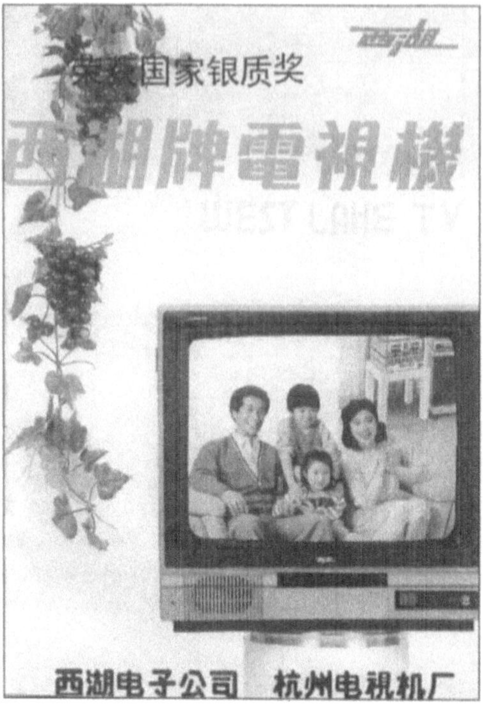

Figure 2.8. This is an advertisement for West Lake TV.
Source: Popular TV, 1989, no. 3.

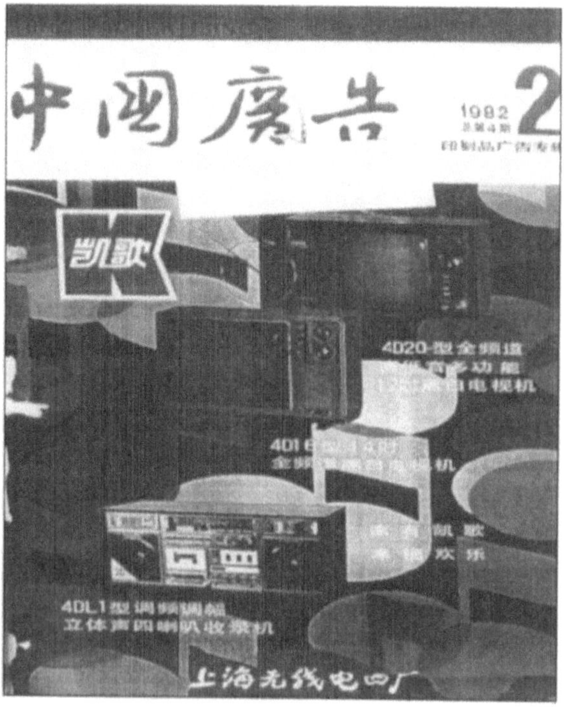

Figure 2.10. This is an advertisement for television sets and radio stereos, both products of the Shanghai No.4 Radio Factory.
Source: Chinese Advertisements, 1982, no. 2.

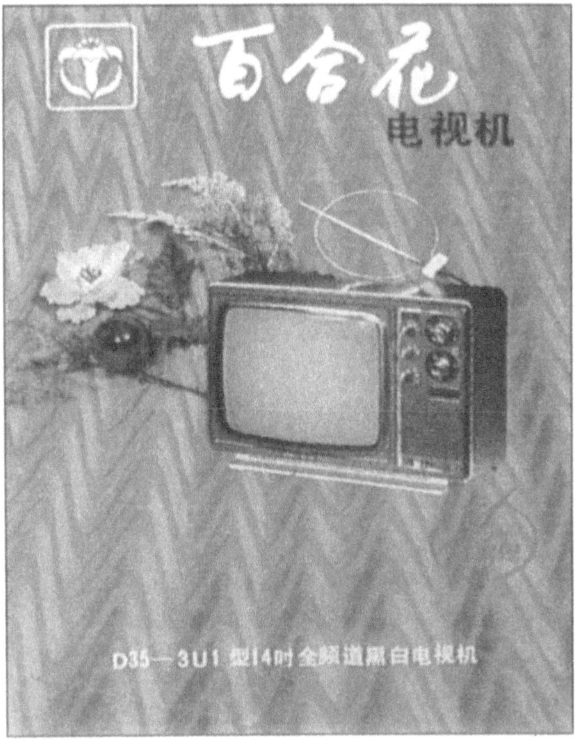

Figure 3.1. This is an advertisement for Lily TV.
Source: Chinese Advertisements, 1985, no. 3.

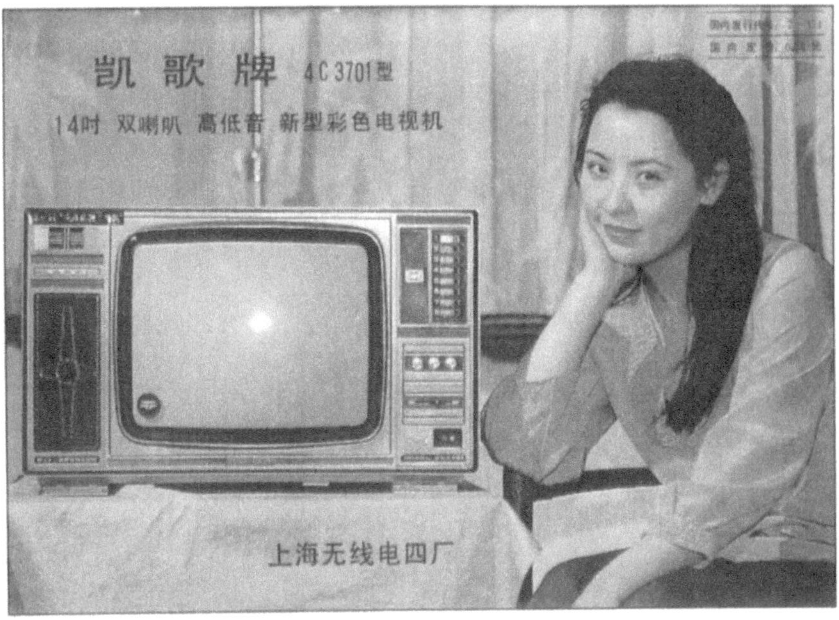

Figue 5.1. A Kaige TV advertisement.
Source: Popular TV, 1985, no. 8.

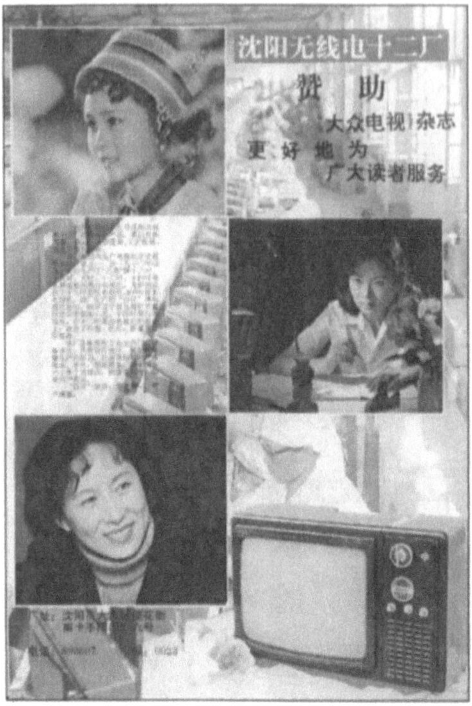

Figure 5.2. A Baihua TV advertisement.
Source: Popular TV, 1983, no. 2.

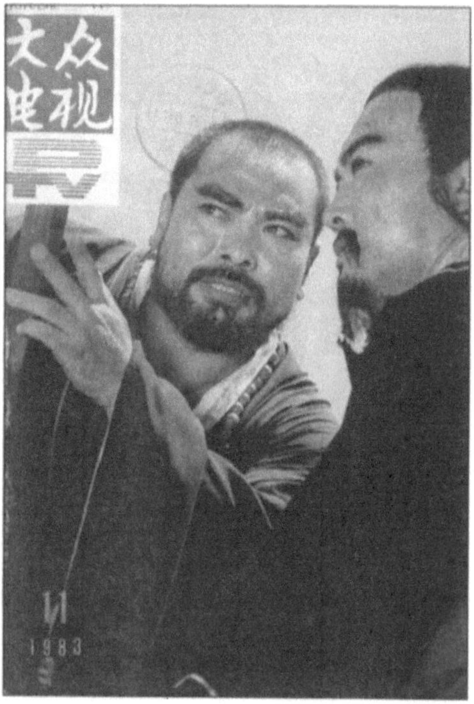

Figure 5.3. The man on the left is a monk.
Source: Popular TV, 1983, no. 11.

CHAPTER 4
TV AND THE OTHER MODERN MASS MEDIA

The relationship between television and other modern mass media is too complicated to examine fully. However, given the importance of this relationship, it is impossible to talk about television without briefly discussing its connections with other media.

Television can be considered from many aspects: technological, cultural, artistic, institutional, industrial, and so on. I could address several of the aspects of TV without drawing a clear boundary between TV and other media. To draw a clear picture of television within its "media family" in 1980s Chinese society, I examine the important cultural traditions that have been created by older media, such as film, literature, opera, and magazines, for the creation of which political and economic power has always been central.

The Relationship among Media

The relationships among modern mass media have been examined by many scholars, Marshall McLuhan being one of the most notable. McLuhan categorized media into two types: "cool" and "hot." According to McLuhan, "cool" media, such as TV, do not require intense audience concentration and leave many blank spaces for the audience to fill in. Film, in contrast, is a "hot" medium, because it requires the full engagement of one sense such as vision, it is intense, and it requires audience members' full attention. Hot media, because of their intense visual impact, do not leave people much room to participate in creating their own understanding of the media. McLuhan also argued that "the content of any medium is always another medium" (McLuhan and Frank, 1995, p. 151). McLuhan's focus on the technological characteristics of different media genres has evoked criticism of his "technological determinism." Canadian scholar Harold Innis has been criticized for the same reason. Innis (1951) categorizes media into two types: "time-biased" and "space-biased." Because different media preserve content and communicate in different ways, Innis argues, every medium carries information either by passing a time channel or by overcoming a geographic blockage. Some media can bind time and space at the same time. For example, stone inscriptions can bind time but not space because it is very hard to move them; papers can bind space but lack the duration to overcome time: they can be burned, destroyed by rain, or lost during transit. Print media bind space but are not as dependable as stone inscriptions in preserving information over a long historical period. Innis's argument illustrates why it is necessary for human beings to have many kinds of media, to preserve and transform information while overcoming the difficulties caused by time and space. As Innis (1951) and McLuhan (1995) argued, human society, to achieve the best communication, must struggle to overcome the weakness of each medium. Their studies, assuming the necessity of each medium, explain the co-existence of all the media human beings have in modern society, no matter whether they are old or new, visual or acoustic.

Technological determinism has been criticized by cultural studies scholars who claim that human agency is always playing with invisible hands to regulate media. These invisible hands are much more powerful than the technology itself. Working from this perspective, contemporary media scholar Susan Douglas examined American radio history and pointed out that radio's biggest influence on American culture has been that it created a daily habit for Americans—to listen (Douglas, 2004). The habit of listening has probably been more important than the specific things audiences listen to, because the habit laid the foundation for the development of several media and related inventions, such as electronic listening appliances (for example, MP3 players). Which media people listen to and why and how they do so are matters constructed by invisible hands, not just by the technological characteristics of media. In other words, authorities and institutions are very important in influencing the types of media we use and how we use them.

Michele Hilmes (2007) has asserted that the competition between different media genres and the media industry's need to achieve variety greatly affect which media succeed with audiences. One of the best examples, according to Hilmes's detailed examination, is how the film industry made an effort, beginning in the 1950s, to engage unconventional content when television started to dominate Americans' leisure time. This strategy was intended to help films compete with family-oriented television programs, which are highly censored to conform with conventional values. Hilmes's argument helps explain why, since the 1960s, the content of movies has increasingly included sexuality and violence. Along with the chaotic social environment in the United States in the 1960s, the competition between film and television has led to these media having distinctive types of content. James Baughman offers a similar insight:

> Television's marginalization of moviemaking, like that of radio, had its advantages [for the film industry]. With most adults rarely going to the movies, filmmakers could be much more realistic or exploitative about sexuality than ever before. The guardians, who earlier would have tried to shut down the exhibitor of the offending feature, had either died or were at home watching television. (Baughman, 2006, p. 200)

The content of many Chinese films produced in the late 1980s supports Baughman's argument. Chinese films had never before contained so many sexual scenes and exposed so much cultural darkness.

As the above scholars have argued, media interact with each other in a variety of ways. It is impossible to understand one medium without looking at the overall media and social environment. By comparing television in 1980s China with television in the United States in the 1950s and the 1960s, this chapter sheds light on TV in the Chinese context.

TV as Projector

Life out of 8 Hours (1982, no. 5) published a story about how a man carried his TV set to different villages at night. The story illustrates a man's unique way of sharing his TV set with neighbors by adapting the government's earlier strategy regarding films. The government took films to villages as part of a strategy to bring high-quality and politically correct entertainment to the rural areas. Many villagers have fond memories of the days when they watched movies played by a projectionist who traveled from village to village. In the interviews and the stories that have been published in magazines, there are several interesting stories about these projectionists who tried their best to serve people and meet the demands of the villages wanting them to visit. Carrying a movie projector and a reel of film, the projectionists attempted to bring happiness to the distant villages by giving people the opportunity to get together to have fun and enjoy the viewing experience. The act of carrying a TV set on the back of a bicycle to make it possible for villagers to watch television programs was obviously based on this practice with films. Similar phenomena occurred during the early stages of television in the United States. One example of a technique developed by film studios during the early days of broadcast television was "theater television, a technique for broadcasting signals onto movie screens in theaters" (Hilmes, 2007, p. 148). Although this concept might now sound unusual, it "gained some popularity early on—when few homes owned television sets—especially for big-ticket events like national sporting matches, but faded due to the FCC's refusal to grant permission for microwave transmission" (ibid.). Apparently the idea of *theater television* shows a rivalry between older and newer media initiating the governmental policies on media. The case in China, which we can call *village television,* is also connected with the way that people live with older media and the political policies, but the ideology is distinguished between American theater television and Chinese village television.

People who had become rich showed television programs in villages to help villagers do the same. The people who volunteered for this work felt they had a responsibility to share their assets with others and repay the government for all it had given them. This was portrayed in magazines as an act of kindness and morality. This relationship between governmental policies and personal wealth was unique to the Economic Reform policy and "socialism with Chinese characteristics." While kindness and high moral character are generally regarded as personal characteristics in contemporary China, during the Economic Reform, the mass media and the state presented kindness and high moral character as resulting from a combination of wise governmental policy and socialist collective spirit. Television, because of its symbolic representation of modern technology, visual-medium characteristics, and capability of being either privately owned or publicly shared, was adopted for the same use cinema had previously been put: to express an ideology of high moral character and personal morality and wise governmental policies. The transition between sharing property and privatizing property was embodied in the TV-watching experience.

Representing the achievement of villages, units and other local communities and the personal achievement and morality (see chapters 1 and 2), TV became one of the successes of Economic Reform most often mentioned by other media in 1980s China. The unique way in which people watched television in China in the 1980s revealed the influence of the older visual medium upon the new and the social ideology underlying both. Showing movies and showing TV both imitate the experience of viewing even older media, such as local operas, and reflect the dominant ideology of sharing.

The War between TV and Film

As was the case with television in many other countries, Chinese television had a limited number of programs in its early stages and so needed to broadcast films. In the United States during the 1940s and 1950s, Hollywood tried to get a foothold in the television industry by producing films for television. At the same time, American television networks tried very hard to avoid the involvement of Hollywood in their programming. China did not have competition between the film and TV industries but did have to face complaints about rapidly decreasing box office sales when TV-set ownership began to increase rapidly in the 1980s. Film producers and officers complained to the Cultural Department that too many films were being shown on television, which was hurting cinema revenues in the 1980s, and asked the government to limit the quantity. The government did impose some restrictions, but television broadcasters continued to show films frequently to fill in their time slots.

In the United States, film had to do something to compete with television and protect its status to secure its own viewership. But the situation in China in the 1980s was different. Although color TV sets were already very popular in the United States in the 1980s, most Chinese audiences were only able to afford black and white TV sets. Therefore, in the 1980s, film still had the competitive advantage in China because of its bright color along with the wide screen and superior sound effects. In the United States, to compete with television, filmmakers initiated many reforms, such as more depiction of controversial material, including sex, drugs, violence, and homosexuality, as mentioned above. In the early 1980s, Chinese media could not do this, because this type of content was not allowed on any kind of media. But the content of Chinese film in the 1980s was nevertheless significantly different from the content of television dramas. While television dramas often embraced the hope of the future and celebrated the changes initiated by Economic Reform, films were going in an opposite direction. The films directed by fourth-generation directors often conveyed a melancholy tone and general nostalgia, while fifth-generation directors[1] expressed an extremely strong willingness to deny the past and destroy patriarchal values of the pre-socialist era.

Many scholars have analyzed the films directed by fifth-generation directors because of their increasing exposure at international film festivals and the artistic and ideological richness of their rebellious narration. Fifth-generation film directors touched on many sensitive issues, such as sexuality, although in a way that was only partially progressive, celebrating primitive masculinity through the sacrifice of women. The fourth-generation directors were more concerned with the glorious days before the Cultural Revolution. They liked to narrate stories in a very poetic tone. Traditional views of history and nature are expressed in their films. People die and seasons change, but life continues and the people grow from the pain and happiness they experience. These films celebrate humanity's ability to face the unpredictable changes brought by harsh nature and cruel politics. Compared to the films of the fifth generation, the fourth generation's films are rich in their poetic pictures. They were not rebellious like the fifth generation. Although scholars have discussed how the fifth generation used stories to subtly express their criticism of the contemporary Chinese government, it is a fact that these directors had been educated by that government, and their rebellious energy was very similar to that of the Red Guards of the Cultural Revolution. They celebrated the power of revolution over political mistakes and called for a masculinity that people had lost under the oppressions of socialist government. Zha Jianying's (1995) interviews with fifth-generation filmmakers reveal several conflicting characteristics that exist in their personal lives and their films: the passion and reality. Influenced by Western directors, the young filmmakers wanted to produce something different and express their view on film as an art, rather than a mere tool of propaganda. They wanted to "go to the world (gain international fame)" (Zha, 1995).

Television was immersed in the complicated and even confusing consumer culture in China in the 1980s. By examining television, one can learn much about the media's struggle to understand what role it should play in people's lives under China during the Economic Reform. Television came into common people's lives right at the moment of transition in China between a pure socialist system and the early market economy. This political context created controversies for and around television. The Chinese government tried to make television a serious artistic and cultural domestic medium; however, the shortage of domestic television programs and the necessity to protect the film viewership made the importation of foreign television programs unavoidable.

Television broadcasters could not survive by depending on film and the other traditional art forms to fill time slots. The Film Department was requiring the Cultural Department to reduce the amount of films shown on TV to protect film production and funding. According to the research of Hong Junhao (1998),

> since the reform, beginning in the late 1970s, the film industry has become more and more reluctant to dispense free movies. Now, not only do films charge a fee for television use, but also a film's price takes into account its cost and the box-office receipts. Broadcasters have been told that they may not televise a film for at least six months after its release, so as not to hurt the box office. (pp. 81-82)

Film producers were unhappy with people's television fever and felt it threatened their own older, better-established visual form. The Chinese social system was more efficient than that of most countries in resolving the competition between film and television. Unlike in the United States, where Hollywood has always been interested in seeking a foothold in any newer visual medium, and was able to gain an important position in television right from the beginning, in China the Chinese Film Department was successful in getting the government to make the appropriate allocation and force television broadcasters to look for their own "food." Therefore, international TV shows began to flood into China. As Hong Junhao has pointed out, "As the trend continued, by the middle of the 1980s most big regional and city television stations had authority to import foreign television programs" (Hong, 1998, p. 60). When domestically produced shows were not sufficient to fill the time slots, even local TV stations became eager to import programming to avoid repeatedly showing the content of CCTV.

One of the best examples is Shanghai TV. As Hong also conveys, "During the early years of the 1980s, by taking advantage of the city's economic strength and broad international contacts, STV imported more programming than the national network, CCTV" (Hong, 1998, p. 61). Consequently, television shows from the United States and Hong Kong were widely broadcast in China. They became very popular because their stories were much more entertaining than those of the domestic TV shows (and those imported from other places). Thus, while film producers sought to go abroad and seek the recognition of the international art film circle, Chinese television broadcasters were eager to invite in foreign programs.

The war between film and TV shows was short and resolved easily by the government. The government's policy to stop television stations from playing movies cannot be considered the primary evidence to explain why China imported so many foreign TV shows in the 1980s. However, it did explain the shortage of TV shows at that time and historical context in which China decided to import the "cooked food" from outside. For many local Chinese television stations at the time, importing shows was an efficient use of resources and money. TV stations could easily pay for the imported shows through advertisements from sponsors. The American and Hong Kong TV shows were welcomed by Chinese audiences, who were attracted by the different narrative styles and the exciting visual techniques in these commercial-driven shows. It is not hard to understand why these foreign TV shows were made affordable and available for many Chinese stations to import in the 1980s. The rapid media production in the United States, which began under the deregulation of media and economic policies of the Reagan government, provided a rich resource that was available for other nations. As Hilmes (2007) argues:

> The U.S. domestic media market is so large that U.S. companies can afford to sell their television series abroad for prices well below what it would take to produce a program of equal quality in most of the smaller nations. Thus, since the 1960s, U.S. TV programs abroad have been inexpensive, of good technical and visual quality, and aggressively marketed by major media corporations. It was hard for newly minted television channels in other nations, struggling to compete and to fill their hours with attractive programs, to resist the lure of American TV. And U.S. shows are popular; audiences around the world enjoy American television series and films, though not always the ones that Americans themselves find most critically excellent or appealing. (p. 292)

Chinese TV had a hunger for content but was short of food, unlike in the United States, where there was a long history of adapting performances of older media, such as vaudeville, theater performances, and radio, to newer media. Because of the consideration of different markets required by commercial broadcasting, which needed lots of diverse programs, U.S. media have always been in the process of looking for new possibilities based on the techniques and performances of "older" media genres. China did have many traditional performing arts, but none of them were integrated into TV shows or production.

The relationship between the U.S. film and television industries is an example of the newer media's adaption of older media's contents. Hollywood lent its rich experiences of visual media production to the TV industry. This trend reached its peak when cable TV gained a foothold in the late 1970s and early 1980s. However, in China the film and TV departments within the government remained clearly distinct, making a discussion about whether film producers should participate in TV production irrelevant. It was the same case with the other arts. Therefore, although China did have many traditional performing arts, there were not many substantial cooperations between any two of them. Shows depicting the traditional arts were shown on Chinese television, but there was no shared production between the makers of the traditional-arts shows and the television producers. Guo Zhenzhi (1997) points out the difficulty that Chinese television faced at its inception when both technology and content were underdeveloped, and the TV broadcasters had to find some "indigenous and primitive methods" to make up for the shortage of both technological tools and TV shows. One of the most striking examples is that of an engineer who used a video camera to broadcast images projected on a wall by a film projector on TV (Guo 1997, p. 132). That is one example of how people overcame technological difficulties through basic technical "cooperation" between film and the new medium, TV broadcasting. Later, when people were able to convert film into a TV format and play it on TV with a much better visual quality than was possible by broadcasting the images projected on a wall by a film projector, many films were aired on TV, resulting in complaints from the film department.

In the early 1980s, both new films and old films were very important for the profits of film companies. At that time, people did not go to movie theaters only to see new movies. Old movies could play for years and would still make money. The phenomena of rapid production of films and constantly changing playlists did not develop until the onset of modern movie theaters. The appearance of the VHS, VCD, and DVD has made the protection of old films not as necessary as it was in the early 1980s, because old movies could be played at home.

The division between television and film was only one of the many reasons that led to the importation of foreign TV programs in the 1980s. Under Economic Reform, TV stations were responsible for their own survival and development. Economic ends motivated TV stations to discover new methods of keeping their viewers interested. These stations also respond to the governmental policies on the reform and the role of TV as an important symbol of modernization and "the throat and tongue" of the government. This potential economic and political gain drove the local stations and the CCTV to find their own ways of importing TV programs that were appropriate for the local and central cultural contexts. The comparatively looser control on the ideological content of television opened a door for more imported TV shows.

In 1989, Chinese Central TV and the Chinese Film Distribution and Exhibition Corporation signed a contract in which they agreed that TV stations could not broadcast new films for two years (Guo, 1997). The war between TV and film was resolved peacefully under the allocation managed by the government and the importation of TV programs from the outside.

TV and 'Its' Print Media

The interaction between television and print media, as between television and film, in China in the 1980s reflects the special role of television in Chinese society. Newspapers and magazines often printed news about television, but this content consisted mostly of contests in which people could win TV sets, articles about how many people had TV sets and articles about how television symbolized the decreasing morality associated with Economic Reform, during which people's material desires were increasing. Television was a symbol of hard work and high morals when it was used in the right way, but a symbol of moral corruption and capitalist spiritual pollution when not. Although television was a newcomer to the media world, it took on the values of that world. In the 1980s, people were concerned with how to use this new medium and how to integrate it as an instrument of modernization. Television itself became a topic that was covered by the well-established print media.

As the amount of programming increased, the need arose to keep audiences aware of upcoming broadcasting schedules. People could use newspapers to plan their daily viewing schedules. Therefore, multiple newspapers and magazines about TV were initiated in the 1980s. *Chinese TV Newspaper* had the most subscribers. *Popular TV* also gained a very decent readership. This type of periodical not only enabled people to easily plan their viewing schedules but also provided background information on various facets of television production, including the personal lives of the TV producers, actors, and actresses. These newspapers also told the audiences what would happen in the later episodes of the TV serials, which made it possible for viewers to continue to follow the plot line of a show even when they had to miss an episode. Some local-TV newspapers also contained recipes and other types of information that seemed to have nothing to do with television. The newspaper not only connected audiences with people and stories they might see on TV but also cast television as a member of the family and a part of people's daily lives as company during dinner or knitting time. When seen in this light, the fact that TV newspapers contained recipes or knitting instructions does not seem strange. The local-TV newspapers also sometimes contained articles about actors and actresses who happened to be from the local area, thus helping the audiences to feel more connected to those actors' shows. These newspapers and magazines were very profitable because of the high rates they charged for subscriptions and advertisements. Although television was often criticized in print media, it created opportunities for the development of new publications.

TV and print media also worked together to teach Chinese people how to consume. The magazine *Life out of 8 hours* published several articles about how to use money efficiently. One article, for example, included a statistical analysis of the necessity for owning various commodities. Although the author concluded that a TV set was not the most essential electronic appliance for a family to have, it was on the shopping list of most families. TV was introduced to people by print media, especially magazines, as an essential item to have as soon as they could afford it. Magazines were serving the readers as a tool of learning to consume with the limited incomes in modern life. Once China initiated Economic Reform, many commodities were introduced into people's lives. After a decade of planned production and allocation, Chinese citizens were receiving different incomes and could begin to design their own consumption structures. Television as an instrument of materialism and education became one of the most important items on people's shopping lists. However, according to most magazine articles on the topic, no matter the salary and practical needs of a family, a TV set was not more urgent or more necessary than other items, although magazines would include TV advertisements in the same issues as their TV criticism. The conflict between serving people and serving economic development struggled with each other in magazines' descriptions of TV sets.

The relationship in China between television and print media was ambivalent, as it was in many cultures between the "old" and the "new," and the older media always had the privilege of criticizing or promoting the newer media. The newer media, on the other hand, created new needs in the older media. The surprising number of new TV-focused magazines and newspapers was evidence of the symbiotic relationship between the TV and other print media.

While TV periodicals were very popular because of their low cost and useful content, they also grounded the collective imagination of provinces, cities, and the nation. TV stations were designed along the model of the government's administrative division, which made the whole picture of TV broadcasting very clear. Responsibility for providing information about domestic and foreign events was clearly divided among Chinese Central TV (CCTV), provincial TV, city stations, and, eventually, county and town community closed-circuit television. CCTV had the most powerful control over news and sports programs. Local television stations were more focused on entertainment or shows having to do with daily life. Beginning in the 1980s, city TV stations worked hard to provide entertaining programs.

An interesting aspect of television broadcasting in China in the 1980s was that people in comparatively remote places often could watch CCTV with a very simple antenna while not being able to access local TV stations. CCTV used a very strong signal so as to be able to keep audiences aware of national news and polices. Because the central government's policies had more effect on people's lives than did local governmental policies, local news and information were not as important as the national news and information. As it has always been in modern Chinese culture, in the 1980s the nation was the most important source of identity. The reinforcement of this sensibility, including by television news broadcasting, is central to the dominant ideology of nationalism.

The political concern of national TV newspapers is strong and clear, as is the content of national TV stations. Therefore, people received political information from CCTV, but relied on local television stations for the more relaxing and much less serious content, such as entertainment. There was no reason to watch the TV stations of other cities in addition to one's own, especially provinces, because all local TV stations repeated the news from CCTV and the entertainment programs of each were highly specific to each region.

The different content of local and national TV stations was very interesting because it was so naturally and logically organized in a vertical administrative system. On the one hand, this system satisfied the needs of diverse local cultures, and, on the other hand, promoted the unity of Chinese culture as a whole.

As a consequence, local-TV newspapers tended to deal with information such as programming schedules, while the national TV newspaper devoted itself more to cultural policies and administrative information about television broadcasting.[2]

Criticism of TV in Magazines and Newspapers

The production of local TV plays helped to redress the shortage of TV programs. Magazines became the most important medium for informing people about the rewards given by TV factories that TV stations could earn by producing plays. Advertisements for TV sets branded "The Great Wall" announced that the company would give a TV set as a reward to any TV station that could produce a certain amount of TV dramas.[3] As mentioned earlier, the increasing popularity of TV sets motivated the creation of many periodicals devoted to TV. These print media, which were dependent on TV, at the same time, supervised the representations and ideology given in TV shows. Two of the most important and well-known national awards for TV programs, producers, and actors were created by magazines. These honors are given out on the basis of a reader opinion poll. Magazine readers, by voting for their favorite TV shows, actors and actresses, participate in the activities of reading magazines and watching TV programs at the same time. However, unlike the synergy among media in the United States, in China, magazine publishers and TV stations were divided by a very clear line based on the different branches of government. Journals and electronic media belong to two different domains in the administrative system. Although the TV magazines and newspapers are based on television, they do not come under the regulations of TV stations or TV industry. This conjoint but at the same time clearly divided relationship between TV and its spin-off print media made supervision by the Chinese Cultural Department on TV very convenient. This is also the same situation as the film and film magazines.

The undeclared but real relationship between print media and TV vested the print media with the dual tasks of introducing and praising television programs and people but also criticizing television as a flawed medium. Television is a visual medium that requires a relatively higher degree of physical engagement from the audience than does radio. Television uses a relatively lower degree of intellectual engagement than books. Its special symbolic role in the process of modernization made it an important target for essays, stories, and comics published in magazines and newspaper. *Popular Cinema* often published cartoons and articles that criticized TV. One of the cartoons (no. 3, 1980) describes a creator of TV programs asking his colleague, "Why don't you go home?" Standing inside of the TV studio, the person answers, "I don't want to go home. This [faked] room is much better than mine." A column, "TV Drama Page," discussed contemporary TV dramas in the 1985 issues of *Popular Cinema*. The satire and criticism of TV in film magazines, unsurprisingly, focused on establishing television's role in corrupting traditional morality and making people materialistic.

Both the realism in film and on early TV programs was not criticized. Because, to some degree, film and literature both were regarded as helping to save the country by educating people in the early twentieth century, especially during the 1930s and 1940s when the leftists were very active. At the same time, because realism was a dominant ideological tool in literature and culture in early-twentieth-century China and then in the People's Republic of China, people criticized TV with the view of realism in the early 1980s. As John Fiske and John Hartley (2003) argue, "the more 'realistic' a programme is thought to be, the more trusted, enjoyable—and therefore the more popular—it becomes" (p. 128). However, "realism too is an artificial construct" (ibid.). Chinese intellectuals assumed there was a reality that TV could actually reflect and display, yet the realism proffered was always under the direction of the party and adhered to proletarian artistic criteria. But popular audiences were looking for this "constructed" reality on TV. In the early 1980s, Chinese TV producers had to be careful to follow the tradition of realism that by then was well established in literature and culture. A TV program could easily fail due to a lack of realism in the early 1980s. Chinese TV producers presented the great accomplishments and the unbelievable sacrifices that the communists had made, as well as the big hopes for Economic Reform. This drive toward realism also led to a propensity of TV documentaries, as we have seen in a previous chapter. The fact that realism was central to television critics was because of the realist tradition established in early-twentieth-century literature and reinforced under communist ideology. At the same time, this criticism also reflects the lack of alternative views on television and the realization that TV served a different role than other media. Although in other parts of the world, TV criticism had, by the 1980s, already developed its own set of theories, Chinese TV criticism had yet to establish its own unique perspective. But before that position was clear, the critical tradition that had accompanied the development of literature and all related media, including print media and radio, was the only tool with which people could analyze television.

The fact that TV critics and literary critics used the same theories to approach their targets created a very interesting relationship between television and print media in 1980s China. Unlike the "strained" relationship between television and magazines in the United States, Chinese print media were seen as having the power to legitimate and evaluate television. This process of legitimacy is reflected in the two national television awards that were established in the very beginning of the 1980s, which have since become the most influential television awards in China. The two awards are "Fei Tian" (Apsaras Awards for Chinese TV Dramas) and "Jin Ying" (Golden Eagle Awards for Chinese TV Programs). Fei Tian is the highest award given to TV dramas by the Chinese government. The judging panel consists of successful TV artists, experts, and people who are important in the TV community. The awards were hosted by a team consisting of representatives from two magazines and one newspaper, *TV Literature and Art*, *Chinese Broadcasting and TV*, and *TV Weekly*. In the 1980s, Fei Tian was awarded annually. The Central Broadcasting Department gave hosting rights to the three print media. Jin Ying is the only nationwide award based on votes from audiences. A Jin Ying was awarded for almost every branch of television programming and production.

The institution in charge of the evaluation of Jin Ying is the magazine *Popular TV*. The procedure for choosing winners has several steps. A local (provincial and municipal) committee of TV artists and experts first nominates several

TV programs, the list of which they send to *Popular TV*. The lists of the TV shows that were published in *Popular TV* include the names of all the nominated TV programs and producers, thereby providing clear information by which the audiences can review and compare what they have watched. They then choose the shows and programs that they think deserve the awards. Based on the votes from audiences and also the readers of *Popular TV*, the panel of expert judges chooses the winners. The close association between the print media and TV is very clearly reflected in the procedure of these two most important national TV awards.

Print Media as the Supervisor and Supporter of TV

As John Fiske (2003) points out, compared with print media, television is more like an oral medium. It functions in a way similar to the way people talk to each other. Because of its oral characteristics, both producers and audiences forgive imperfection in the flow of television. We cannot go back to check those contents we have just watched on TV and we do not even want to go back to check, which is the same ethic we apply when talking to and listening to people in our daily lives. McLuhan's (1995) explanation of hot and cold media can also help us understand the characteristics of television at this point. Skipping content presented on TV might be because television is a cool medium, and audience tends to ignore many things on it. Tolerance of the imperfections of television is one reason many intellectuals do not think of television programs as a serious art form, like movies. The constant flow of television also causes the ignorance of audiences. It is not hard to remember a movie because the content is well organized and short. However, TV programs, because of their flow, can easily escape people's attention. Consequently, print media supplant this "flaw" of television.

TV magazines and newspapers provide concrete records to help people review and remember the television programs they have watched. Although television viewing may have eaten into much of the time that people might have spent reading magazines or newspapers, the flourishing of print media in the United States in the 1950s and in China in the 1980s reveals that television, paradoxically, played a role in the fast development of modern print media rather than threatening them as television did to films. Television can display more information than print media can, but television depends on print media to provide audiences a space in which to step back from the flow. In this sense, television is a different medium than print media, not a replacement or improvement on the older medium.

As Arvind Rajagopal (2006) has argued, "The world of print media has often received nostalgic treatment, as an era of rationality that monopoly capital and consumerism have torn us away from" (p. 285). Print media, in contrast to oral media, have the feature of normalization and a cultural sanction as a "rational power." This is because "print extruded the authority that previously was secret and located in privileged bodies, and rendered it abstract and diffusible" (ibid.). Thus, the "rational power" of print media, such as magazines and newspapers, naturally functioned in a surveillance role over the "newer" electronic medium, television.

The relationship between TV and print media in 1980s China is further illuminated by considering the history of film awards. The first national film award, Bai Hua (Hundreds of Flowers), was created in the 1960s, and it was hosted by the magazine *Dazhong Dianying* (*Popular Cinema*). The name of the magazine reveals the relationship between magazines and film: it is a magazine born by film. Readers mailed their votes for various films into the magazine, and the magazine editing office took care of the evaluation process, creating a panel of judges. Apparently, the relationship between film and print media had also experienced this supervisee and supervisor process. But in 1981, when the other national film award, Jin Ji (Golden Rooster), was created, the National Film Association had already developed its own authority and had its own system to create a panel of judges and organize the evaluation process. Afterwards there was no involvement of the print media in this process. This situation was indicative of the historic interaction between print and electronic media. It also helps us understand how film and TV, both combined visual and acoustic media, have been given vastly different cultural roles: the former is related to the authorities from art institutions, and the latter requires more audiences' participation.

In the 1980s, Chinese journalism education had already established a well-organized system. There were many qualified programs to train journalists, editors, reporters and relevant professionals. Chinese journalism went through big changes over the course of modern Chinese history, and journalists had always been respected by Chinese people because of the loyal role they played in helping the population stay up to date with governmental policies and teaching them to love and support the government. During the early years of the twentieth century, rebellious communist journalists fought for the common people and disclosed the abuses of the pre-communist government. Journalism has always played an important role in the revolution led by the Chinese Communist Party. Therefore, nationally owned magazines and newspapers are given a glamorous record in communist histories and already created a professional and well-recognized tradition and educational system. Print media, therefore, had earned great respect before television arrived. Compared to print media, television was a newcomer in 1980s China. Television served more of a technological than artistic or literary function in the education of the populace. TV was the symbol of modern technology and lifestyle. Therefore, the experts and artists of print media were more than qualified to be the judges of the television awards. As a result, the fact that magazines and newspapers were given the right to serve as arbiters of television from both a technological and cultural perspective seemed very natural in the 1980s. The role that print media played in legitimating television was further illuminated in the 1990s when audiences began to cast their votes through the telephone or Internet rather than mailing them into the publishing offices of magazines or newspapers. The history of the interaction between TV and other mass media can give us a better understanding of how the role of different media genres have been evolving since the birth of each of them.

Notes

1. China reopened its colleges and universities and resumed a national entrance examination system in 1977, at the end of the Cultural Revolution. "Fifth generation" refers chronologically and aesthetically to young filmmakers who graduated from the Beijing Film Academy in 1982 (Cui, 2003, p. 100).

2. Electronic media, especially the Internet, have seriously threatened the well-established system of fostering Chinese nationalism because it conveys information from the outside that cannot be censored, but, at the same time, it strengthens nationalism through its powerful nationalistic broadcasting system.

3. See the advertisement for The Great Wall TV published in *Popular TV*, 1982, 5

CHAPTER 5
PROVIDING MORE CHOICES?: DOMESTIC AND IMPORTED TV DRAMAS AND GENDER CONFUSION

Opposed to the early-twentieth-century Chinese women, who were fighting for their rights, Chinese women, especially intellectual women in the 1980s, were struggling with the concept of what it meant to be a woman. Many Chinese women writers of the 1980s were crying out with questions like: Where is masculinity? What is love? And who am I? Their confusion was embedded in their stories. Their dream of equality between the sexes was still imperfectly realized after three chaotic decades of struggle. Like the policymakers in the 1980s who were thinking about how to effect change, Chinese people started rethinking the meaning of life and what a person should be in life. Gender was one of the most important topics that people, especially women, talked about as they contemplated the immediate past and their rethought contemporary and future life.

Television provided a screen on which members of the public could watch the dramatic playing out of the contentions of these intellectuals. At the same time, as a symbol of a modern and a better life, TV sets were portrayed with certain ideology in the cultural context that emphasized the longing for modernization and economic prosperity. In television's roles as a communicator of the nation's dominant ideology and a technological product, what television was depicting on its screen and how TV sets were being promoted in commercials reflect the complex struggles faced by Chinese people in the 1980s. If television provided an image of gender in other cultural contexts and an idealized template of gender norms, this ideal resonated with pre-existing gender notions. However, the commercials promote consumption of TV sets by presenting their benefits in the contemporary moment, the 1980s. The main question this chapter seeks to answer is how gender is related to the imagination of idealized gender norms in and before the 1980s.

Wandering between Boundaries: A Brief Review of TV Dramas Produced in 1980s China

"Beginning in 1981, the PRC (People's Republic of China) began producing its own serials as opposed to single-episode dramas" (Lu, 2000). By the late 1980s, Chinese Central TV Station was producing several TV dramas. According to the records of "Zhongguo Dazhong Dianshi Jinying Jiang" (The Golden Eagle Award for Chinese TV Programs), Chinese TV stations produced several TV dramas throughout the decade. Some of them were based on classical fiction, some were based on the novels of young contemporary writers, and some were new stories about young people employed in modern industries, such as *Da Jiudian (Great Hotel)*. There were also TV dramas praising the People's Liberation Army. Although this range of dramas seems very diverse, it is clear that Chinese TV was still struggling between playing an educational role and playing an entertainment role. "The television dramas that had most appeal during this time were those described as 'drama of pains' (Shanghenju) or 'retrospective dramas' (Fansiju), which recounted the adversities of past political turmoil. The appeal of these dramas was subsequently displaced by 'reform dramas' (Gaigeju), narratives about political and personal conflicts that were occurring during the reform period" (Yin, 2002, p. 31). All of these dramatic genres ultimately served the purpose of educating people on the necessity of economic reform by criticizing the past and examining the present.

Of these genres, *The Snow City* is the most like a "drama of pains"; *The Policemen in Plain Clothes* exemplifies all three genres because of the its chronicle narrative form and *The Bell Tower and the Drum Tower* is close to a "reform drama," but its reform agenda was tempered by description and narration of daily life in the 1980s. Most of the dramas shown on Chinese TV in the 1980s were based on classical writing or fictions. The plot of these TV dramas still depended on literature. They are "dramas adapted from classical and modern literary works" (Yin, 2002, p. 31). *Xiyouji* (Journey to the West), *Honglou Meng* (The Dream of the Red Chamber), and *Sishitongtang* (Four Generations under One Roof) were all adaptations of literature from earlier eras. The first two were written during the late Ming Dynasty (fourteenth-seventeenth century) and the latter was written just before the establishment of the PRC. They belonged to the early Lishi Ju (historical drama). Therefore, their emphasis on patriarchal power and on women as objects of male visual pleasure and objects of the male gaze seemed natural to modern audiences because of being set in a historical time period.

Journey to the West, based on a fairy tale from a novel, was very dramatic and was the first Chinese TV drama to include special effects. *The Dream of the Red Chamber*, based on a story about a woman from a high-ranking family in the late Ming Dynasty, included hundreds of young women characters, beautiful fabrics, ancient decorations, and luxurious interior designs, giving audiences great visual pleasure in spite of its sad story. *Four Generations under One Roof* is about the experiences of a middle-class family in socially chaotic early-twentieth-century China. These shows were far from the daily lives of viewers in the 1980s because the shows' creators rarely changed the content of the original stories. Instead, in line with the realism that dominated Chinese TV in the

1980s, they stayed true to the original story. These television dramas, which were very popular and won many awards, are obviously patriarchal, reflecting historical gender roles. Thus, the construction of gender roles in these TV dramas mostly echoes Chinese authorities' criticism of feudalism's gender divide wherein women are oppressed and men dominant. At the same time, men are suppressed by the political system in the TV shows. Class struggle and nationalism are more important in the ideology of these historical dramas. Like many other historical stories in China, these stories reproduced by the TV broadcasters intended to make viewers feel fortunate to be living in the present. Audiences gained confirmation that their modern lives were better than those of people in feudal or even recent, but pre-PRC, society.

Thus, although the programs sometimes reflected contemporary ideology and contemporary interpretations, the portraits given of masculinity and femininity were not new visions but quite familiar. Women's ambition for power is depicted as being dangerous for themselves and other people. In *The Dream of the Red Chamber*, any femininity or indifference to politics on the part of men causes tragedy to themselves and to their families. These more traditional views on masculinity and femininity in the traditional stories invite comparison with the portraits of men and women in contemporary TV dramas, although the contrast might not be intentional on the part of the TV broadcasters and the contemporary shows' creators. *The Snow City*, *The Policemen in Plain Clothes* and *The Bell Tower and the Drum Tower* were all adapted by contemporary male writers and also directed by men. Compared to the works of women writers in the 1980s, the TV dramas that had male directors display a similar struggle to comprehend "real" masculinity and femininity. But these TV dramas' representation of a gender crisis appears much more patriarchal. Their views on gender are clearly based on their stronger belief in the natural differences between men and women, which, as we all know communism officially denied. The ways in which these TV dramas expose a struggle with gender norms are analyzed in depth later in this chapter.

While the contents on TV display the controversy and struggle over gender norms, imported TV dramas brought different ideas about masculinity and femininity into China in the 1980s. Shows imported from exotic cultures, especially those cultures perceived by the Chinese to be more modern and advanced, were influential in their gender prototypes. The introduction of these foreign views of gender added new complexity to Chinese contemplation of gender norms. Because of the transitions China was making, from socialism to economic reform, and from isolation to involvement, gender issues, which are never clearly explained by any sociological agenda, become one of the most important issues to consider when one studies 1980s Chinese television.

Modernity, Socialism and Gender Roles

In her critical studies, Rita Felski (1995) provided a provocative analysis of the relationship between modernity and studies of gender. Felski's argument sheds light on the central conflict within the Chinese media's representation of gender. Like many scholars, Felski believes that it is very difficult to precisely identify when the "modern" began. Moreover, modernity is experienced differently in different cultures. Being modern does not refer exclusively to life in sixteenth-century Europe and beyond but rather to an attitude that is forward-directed, critical of the present, and sometimes paradoxically reminiscent of the past. In this, Felski echoes Marshall Berman's (1988) examination of modernity. What is the central problem of gender identity in socialist China? As Monique Wittig (1992) has argued, because in Marxism unity was a precondition to achieving power and success, women were not allowed to think of themselves as a separate social group:

> On the practical level, Lenin, *the* party, all the communist parties up to now, including all the most radical political groups, have always reacted to any attempt on the part of women to reflect and form groups based on their own class problem with an accusation of divisiveness. By uniting, we women are dividing the strength of the people. This means that for the Marxists women *belong* either to the bourgeois class or to the proletariat class, in other words, to the men of these classes. (p. 9)

Wittig here points out a common gender identity problem in socialist countries. Recent studies on Chinese women and gender to some degree conform to Wittig's argument. Building on the observation made by both Meng Yue and Dai Jinhua that both men and women in socialist China obeyed the sexless collectivity or its symbol, Cui Shuqin (2003) argues the following:

> The collective moral directive to sacrifice a female self in the name of a social role and to suppress one's gender identity in the name of equality compels Chinese women to shun difference as an aberration and to embrace the formless anonymity of consensus as an ideal. Beyond social and political explanations for the myth of Chinese women's emancipation lie other factors: the lack of gender consciousness by women themselves and the psychological burdens of self-sacrifice. To put everyone else's needs (husband, family, working unit, the state) above her own has been the defining value of a virtuous woman. (p. 56)

Therefore, in traditional culture, Chinese women are self-sacrificing according to their consciousness of themselves as women, but in socialist culture, Chinese women are self-sacrificing without enough consciousness of themselves as women. Moreover, in socialist culture they sacrifice for more people, not only for their husband and children, but also for their working unit and the state, which stand symbolically for the people as a whole. If the modernity in the period following socialism caused Chinese women to rethink their gender identity by criticizing the present and revisiting the past, their consciousness of themselves as women would be the first thing they wanted to gain back (Li, 1994, pp. 306-309). This consciousness of oneself as a woman constitutes the biggest difference between being a woman in pre-Mao society and in the contemporary reality. The female consciousness and the social reality caused new problems concerning gender identity in the 1980s.

As Ravni Thakur (1997) has pointed out, gender norms in post-socialist China "represent a desire on the part of women to revert to older forms of habitus and show the way dominant values work through consensus and acceptance rather than by force" (p. 56). The awkward relationship between Marxism and the liberation of women in socialist China has been exposed as potentially problematic by intellectuals, especially women writers, since the Economic Reform. While the socialist government was criticizing feudalism for its abusive treatment of women, media representations conveyed a strong criticism of the construction of masculinity and femininity in 1980s China and a nostalgia for the more "traditional" gender norms. However, women were already "liberated" from the domestic housewives' life and they definitely wanted to continue their role as productive laborers. Women's concurrent roles as laborers in the domestic arena and as "equal" laborers to men in public working places made their lives difficult. "The new orthodoxy on gender relations is a curious mix of patriarchy and socialism, where the tensions between women's productive and reproductive roles remain unresolved" (Thakur, 1997, p. 62).

At the same time, with the onset of a consciousness of "individuality" and the rejection of the "sexless collectivity," both men and women writers were urgently "searching for the real men" in the 1980s (Lu, 2000; Zhong, 2000; Louie, 2002). The different ideals of masculinity and femininity suggested in the writings of both men and women brought to the surface gender problems that socialism could never resolve but only ignored under the gender equality policy. Between the consciousness of individuality and indoctrination of nationalism, and between the assertion of gender difference and the belief in gender equality, both masculinity and femininity were troubled.

While these competing values remained unresolved in the 1980s, modern gender norms and prototypes were being imported into China from the West. Television, as a medium subject to governmental censorship as well as a stage for popular culture and the reflection of elite, intellectually progressive thought in 1980s China, provides an interface from which we can gain a closer view of the conflict among gender norms caused by the nostalgic past, the problematic present, and the unpredictable future.

Naturalizing Gender: *The Snow City* and a "New" Idealized Femininity and Masculinity

The Snow City was a TV show based on the novel of the same name by Liang Xiaosheng, one of the male writers who best represent the 1980s. Liang Xiaosheng had experienced the movement started in the 1960s in which young people were sent to the countryside and remote areas of China to help "construct the nation and contribute to people." It was part of Mao's plan to "conquer nature," by enlisting all the young people to serve as laborers to develop backward places and at the same time provide them with spiritual and physical exercise. Many young people were excited by the movement and at its outset were very proud of themselves, especially the youth who went to Bei Dahuang in the far northeast as members of the "Jianshe Bingtuan" (the Construction Corps). Because the Construction Corps followed the military system and all the young people wore military uniforms, for the young people it was joining the army although their main assignment was often agricultural works. Serving in the military/the Construction Corps also brought pride and honor to young men and women. They were excited and motivated to the same degree by the military that they were disappointed and lost after experiencing the drudgery of fighting nature in far Northeastern China. They wanted to return home. It was this cultural context on which Liang Xiaosheng's stories were based.

In *The Snow City*, Liang told the stories of many characters and how their lives unfolded after they returned from the countryside to the city in the early 1980s, eleven years after leaving it. The focus of the novel was a girl who was born into the family of a high-ranking official. The makers of the television drama chose to focus instead on Xu Shufang, whose story was one of the book's smaller ones, narrated from the viewpoint of the main character. Xu is a quiet, weak, and attractive girl. The TV drama focuses entirely on what she experiences when she goes back to her home, where her father has committed suicide and her stepmother and stepsister treat her as a burden.

Many aspects of this character raise questions about transitioning gender roles in the 1980s. Xu is made to represent all the young people (who, being close to thirty, were not even very young anymore), both men and women, who went back to cities after the Cultural Revolution. By focusing on Xu's story, the makers of the TV show simultaneously fulfilled two goals: one was to criticize the political power that brought tragedy to the young lives of that generation, and the other was to raise people's sympathy specifically toward the women who were part of this group. Although, according to socialist gender norms, women were the same as men, part of a "sexless collectivity," they finally had to face the impact of gender difference in post-Cultural Revolution China. The TV show's portrait of this character is very multifaceted. In one scene, Xu kneels down to beg for a job from a government official. She often cries, such as when she suffers wrongs or feels afraid of men's threats. She appears to be physically fragile. However, she is given a job shoveling coal at the train station, where all the other workers are men. She gives all her earnings, as a wedding present, to her stepsister, for whom she feels truly happy. She does not have nice clothes and instead wears an old military uniform almost all the time. Her hair is often wet from rain or sweat and she never has a nice hairstyle.

In one scene, Xu Shufang is looking up at the hill where hundreds of men are shoveling coal from a train, and she shouts, "Who wants me? Who wants me?" She is desperate to work and have some income. She wants to have a job even if it means working in a man's world. But the men answer her, "Are you a girl or a married woman? We will think about your quest accordingly." All the men laugh at her. She is very scared and faint as she runs away from the laughter. She had been used to living in a masculine environment in the northeast and was educated to work alongside men to serve the people and the nation. Her shouts for work are like an SOS for help from "comrades." But the response she receives is totally sexual. As the story proceeds, she is saved when four kind men, who work together shoveling coal, take her to work with them. A very young man in the group says, "Old sister, borrowing your words, we take you!" She is fortunate to meet this group of moral men who treat her as a girl but without sexual desire. This episode puts people into three categories, three different ethical positions: immoral men, kind men, and women who believe in living and struggling in a man's world.

The TV adaptation of the story reflects contemporary confusion about gender. It is impossible for viewers not to sympathize with the pale girl, Xu Shufang, as they see her kneel down to beg for a job and watch her doing a dirty and exhausting job with men. She represents both men and women from the Construction Corps. By showing Xu begging, the TV show exposed the bitterness of the Construction Corps workers' lives but at the same time hewed to traditional gender roles: the masculine spirit of Xu's male peers, and the weakness of Xu as a woman that has to face the lack of an appropriate position in the public space.

This portrait, although sympathetic, was constructed through a man's eyes and set in a man's world. Throughout the story, Xu Shufang is saved by men, including her ex-boyfriend, who gives to her his pass back to the city, and her husband, who is the one to lift her off the floor when she faints at the train station. Xu's stepmother, on the other hand, throws her onto the street, rendering her homeless after her stepsister gets married and Xu loses her job at the train station. The stepmother's actions force Xu to ask Qiang to marry her and give her a place to live. Xu must betray her boyfriend in order to marry the man whose kindness she depended on. Xu Shufang, with her often sweating face and wet hair, struggles throughout the drama to be independent and survive life's hardships. Fortunately, as the TV drama portrays, she eventually realizes that the love between herself and her husband is the true love she had been looking for. In her role as a woman coming back from the corps, Xu Shufang was a new model of heroine in the Chinese media: She did not have strong political beliefs or ambitious ideals. All she wanted was a life that she could survive. When all the people surrounding her treated her as an unmarried "old girl" and a woman, she continued to try to live as a good woman who was not using the excuse of being female.

At one point in the story, Xu's husband is told that she betrayed her respectable ex-boyfriend to marry him. He does not talk to her for a long time. Xu Shufang then leaves her home to work in a warehouse, carrying heavy boxes, each almost as heavy as herself. Her husband eventually comes to ask her to return home and says, "I'm a man, and I can't let you do this kind of job." Xu Shufang answers, "I'm a woman, and I need to do this kind of job." She is attempting to prove that she is neither weak nor dependent and can do a man's job as long as she can get one. Struggling with all her might, she can barely walk with the huge box on her back, but she seems to want to use the burdensome job as distraction from her emotional and spiritual suffering. In a society that declared the equality of men and women before asking them to go to the countryside and the corps, Xu does not know how to show she is a respectable woman other than by demonstrating through her actions that she could be the "same" as men. As long as she had a job, Xu Shufang would survive by herself, but when there was not, the best choice for her was to marry a kind man. She makes her choice and then must suffer as a result, because she betrayed her ex-boyfriend, a respectable man. She obviously had not married for money, and she must also demonstrate that she had not married to be taken care of.

The portrait of Xu Shufang reflects the confusion of the author and many intellectuals about the proper role of women. Xu Shufang struggles continually with whether to be weak or to be strong. On the one hand, she wants respect as a worker among men in the corps, but on the other, she accepts her boyfriend's gift and returns to the city. On the one hand, she wants to survive independently but on the other finds she needs to marry her husband to get a home. She was suffering because of the guilt for the two men and her suspicion about her own morality. The only way that she can think of to get rid of her suffering is to engage in physically challenging jobs and compete with men while supporting her husband with her income, thus playing the masculine role.

In this story, the roles of both genders are depicted as being interfered with by the political system's "huge hand." The story reflects a man's consciousness about the difficulty of being a woman in Chinese society in the 1980s. It also displays a patriarchal ideal for woman. Both the book's author and the TV show's director maintained the traditional roles of men and women: men should be saviors to the women they love, while women's role is to be forgiven and sympathized with. At the same time, women should be pure and they should stay with the men to whom they have given their virginity. At one point Xu Shufang decides to return to her boyfriend, but on her way back suddenly realizes and declares out loud about her husband, "I love him."

The story soothes the struggle that it initiated from the beginning; Xu finally makes up her mind to be a good wife to her husband, to whom she gave her virginity. She and her husband both know that she cannot feel comfortable with being a housewife and the economic situation will not allow her to be a housewife, either: she feels the need to be strong and work outside with physically challenging work. However, she does this not because she feels empowered to be independent but because she wants to use her labor on behalf of her husband; it is her way of being a good, virtuous wife. Xu's double role as an independent woman under socialism and a patient and caring wife who is always ready to sacrifice herself for her family members (as she did for stepsister) was fully evident by the drama's end. The portrait of Xu Shufang confirms Thakur's argument:

> Although socialist orthodoxy stresses the importance of women's economic independence and espouses equal opportunities, women in China continue to play their traditional roles, especially within the family. The result of this ambiguity is the grueling double burden faced by women, especially in urban China . . . (p. 144)

The television version of the story denounces the love that Xu Shufang voices and in doing so also denounces a progressive view of love. People in early-twentieth-century China were taught that romantic love was the result of revolution. Love was enabled by the revolution that destroyed the Qing Dynasty, and it was also the result of liberating people to escape arranged marriages so they could choose their own partners. Many men and women participated in the revolution by escaping from arranged marriages. Although the question of what true love is too hard for many people to answer, doing so was very easy during the Communist Revolution, when many people felt that love for another was based on sharing the same ideal—liberation for human beings. This conception of romantic relationships was the predominant view in Chinese movies beginning in the 1950s.

The Snow City denied the idea of unconditional love based on a lofty ideal and also renounced love that was conditional and changeable. The TV drama's attitude toward love sounds like a liberated view for a modern culture in which changes were taking place out of expectation. However, women were rarely allowed to adopt this view without being impugned as immoral. At the same time, woman must know that it was impossible for her to be saved from the difficulty caused by society by women (such as Xu Shufang's stepmother and stepsister). Besides, women's marriage should be based on true love rather than dependence, as Xu Shufang's was, though she only realized it after getting married. The TV show suggested that marriage is conditional and love is conditional. Fortunately, the two men in Xu Shufang's life allowed her to make her own moral choices without applying any force.

The Snow City acknowledges modern China's gender differences and inequality, which were culturally rather than politically constructed. During the Mao era, the differences and inequality had been ignored, by the policy of gender-neutral job distribution in place since the 1950s. The awareness of gender differences was concealed under the dominant ideology that it was the politics and governmental policy of pre-socialist society that caused the mental and material suffering of the people at that moment. If the youth of the period had not gone to work in the northeast, if Xu Shufang had not had a chance to go back to the city, and if she had not been forced to work with men, her marital prospects might have been different. But all those factors existed because of political reasons. People's lives were made trivial under the powerful orders of the government. While women writers were asking where the masculine men were, *The Snow City* showed that they were everywhere. It was not the difference between genders but governmental decisions that caused gender-confusion and suffering of both men and women.

In *The Snow City*, men can always survive because there is always heavy physical work no matter in what kind of society and because women always need men's help. This challenged the idea that men and women were equal, an idea that had been promoted since the 1950s, especially in the Big Leap Forward movement (Da Yuejin). *The Snow City* presents equality as being impossible to attain and acknowledges the differences between men and women. By criticizing the "big hand" of the government, *The Snow City* trivializes the gender issue by treating it as just one of the many consequences of political decisions made in socialist China. Such a description of gender as rooted in the social system "demonstrates that gender relations remain an aspect most resistant to change" (Thakur, 1997, p. 60). When political or military power can destroy a political system and rebuild a new society, many things are easily changed; however, in some areas, especially people's view on gender, the newer ideas always have to struggle with the older norms.

In *The Snow City*, Xu Shufang is a lonely woman living in a man's world in which men are, for the most part, nice to her. Even though her male coworkers make sexual jokes, they do not really intend to violate her. When she is fed up with the jokes, Xu hits the men with a food container at lunch. The men decide to leave "this crazy woman" alone. These men do not exactly look or behave like gentlemen, but they are not immoral. They just think it unusual and funny to watch a woman doing the same job they are doing. Both Xu's eagerness to show that she can do a man's job and the fact that men find this funny reveal that, despite the official, governmental view before the 1980s, men had never really thought women could do any job men could do. There were few opportunities for women to show they were emotionally strong and morally superior other than by showing their tolerance of the unbearable physical challenge of men's work. But as the TV drama version of *The Snow City* represents it, this effort was ridiculous.

This TV drama, written and directed from a man's perspective, displays the struggle with gender identity in a society in which people were still working mainly for food and shelter. By displaying conflict, the TV drama reflected, although not explicitly, a new consciousness about gender and modernization. Xu's misery begins the moment that she returns to the city. She must deal with many challenges, things she did not anticipate although she was born and spent her teenage years in the city. She did not expect there would not be enough jobs for everyone. She did not expect that she could only live with the help of men, despite the fact that the revolution had promised equality among men and women. The narration of the TV drama reveals that politics manipulated people's lives at the time. However, by portraying many kind men, especially the ethical men coming back from the Construction Corps, *The Snow City* imagined a new, idealized society in which men should take care of women and women should be staying with a man despite obstacles unless they do not have any other choice.

The conflicting presentation of a conditional love and relationship, and the unconditional generosity of men to women, and the relationship between marriage and women's virginity, reinforced the traditional Chinese belief in reconstruction of an idealized masculinity and femininity by dominant patriarchal ideology in 1980s China. The idealized masculine men should be generous and caring. Unlike the presentation of masculine men in 1990s films and literature, which reconstructed masculinity as "a projection of primitive vitality," full of libido, the masculinity of the 1980s was "the refined, reserved, effeminate scholar-intellectual type prevalent in Chinese culture" (Lu, 2000, p. 30). And the idealized feminine woman, exemplified by Xu Shufang, should be naturally beautiful, even if donning an old and drab military uniform with sweating hair, as well as loyal to her man.

At the same time, this concept of femininity required women to believe that they should never give up trying to survive on their own even if doing so seemed impossible. Life is hard, but *The Snow City* suggests that if men and women are "good" enough they can always overcome difficulties and live with peace and hope. However, men and women are not equally capable; rather, men are the saviors of the virtuous women. This TV drama's representation of the relations between the genders differs from that of the dramas produced from the 1950s through the 1970s, but it continued the gender representational tradition of novels from thousands of years ago in which heroes were always the saviors of women, and women were supposed to fall in love with their saviors. The difference in this TV drama, however, was that because of the political context, the hero, like the heroine, was powerless. The woman could not receive from him a Cinderella's beautiful clothes and shoes. The TV drama romanticized the bitterness of these common people's lives by combining the traditional ideal of hero and beauty with an idealization of the beauty of morality and spirit.

As already noted, *The Snow City*'s ideal of femininity and masculinity differs from that of socialist China. The drama began by acknowledging people's individual desires and went on to complicate socialism's concept of love as well as the meaning of gender. The show's implicit and explicit expression of gender ideals reveals an increasing affirmation of "individuality." The individuals who came back to the city from the rural areas separated from each other as soon as they got off the train. Although the people had not enjoyed their lives in the rural areas, the feeling of being lost in the city made them nostalgic for the collectivity they had in the past.

In *The Snow City*, the young people's relationship with their closest past was severed and they were set into a new love relationship and the most basic unit of community—family. Life in the rural areas was presented as only a temporary collectivity among people living in the same place and under the same conditions. Food, shelter, love, sexual desire, and marriage became the central themes of city life in the TV drama, and all of these themes provoked characters to think as individuals, in contrast to their socialist education, which had taught them to be unselfish, serving the people and the country. City and countryside were presented as opposites, and life in the city and life in the Construction Corps were distinct, because the former was much more individual than the latter. A clean line was drawn between characters' lives in the rural area and in the city; however, the characters' imagination of the past in the rural area was still conflicting with their contemporary experience.

Although political power was central to *The Snow City*'s critique of the difficulty of a generation's life, the show's presentation of gender roles was based on assumptions about the natural differences between men and women and on the belief that men should play the political role in a patriarchal society. The show's idea of femininity is naturalized at the moment when it evokes sympathy with Xu Shufang's experience. Her job was not a job that a woman should do, suggests the show; society forced her to do it. This ideology was very different than the propaganda of the "iron girl" that the Chinese media exuberantly displayed in the 1950s as well as the images of the revolutionary female soldiers (Hong Wei Bing/Red Guard) that were distributed in the 1960s and 1970s. Xu Shufang represents a woman who was educated in the ideology of equality between men and women and then suffered from this belief after she came back from the corps. Her consciousness of being a woman was awakened into the harshness of the new urban life, illustrated by the moment when she could not move with the heavy box and yelled and cried by herself.

At the same time, in *The Snow City*, there is not a clear sense about the future of all people. The show suggests that its heroes and heroines were given a moral and spiritual treasure in their past experience, when they worked in the corps, and that this morality and spirit would accompany them into the unknowable future. Without the idea of future, the show's combination of the gender norms of socialist discourse and of Chinese tradition imposed a "newer" gender discourse. In this newer discourse, the division between genders was naturalized according to sex differences. Xu Shufang finally finds something she can hold on to—her love for her husband. And she loves him even more when he eventually sacrifices himself and goes with the policemen because he does not want to reveal information about people who caused an uproar at a national exam. At the moment when Xu Shufang watches as her husband is arrested, she starts to cry; her role as a caring wife, a sacrificing woman, and a person trusting of ultimate governmental justice is complete. This female role constructed from a patriarchal viewpoint is finished in a moment of silence and the peace that Xu Shufang has at the conclusion of the show.

Visualizing Women: Who is the Viewer of TV Advertisements?

While the TV dramas, such as *The Snow City*, display the struggle and confusion of gender roles, as the carrier of these dramas, TV sets were promoted in the advertisements with a clear gendered message. Advertisements of the 1980s might seem dull according to modern sensibilities. Viewers nowadays are accustomed to seeing complex visual images. Paradoxically, these "dull" ads that served as simple introductions to particular items provide rich, clear information about their social context. The presentation of gender roles in TV advertisements can help us look at the unspoken assumptions about gender in the modernizing society of 1980s China.

In the 1980s, images of women dominated in ads for TV sets in magazines. In the *Kaige* TV-set ads of 1984 (see Figure 5.1 in Photo Spread), the visual elements were very simple: a closed curtain; a TV set on a table covered with a cream-colored tablecloth; a woman sitting next to a table, facing the camera, inviting a sexual gaze; and some text, Chinese characters spread out on the top and the bottom of the ads giving information about the TV and its manufacturer. It might be hard to find another ad so simple anywhere else. The ad's meaning might be confusing to readers accustomed to a much more complex and protracted visual encoding. The woman in the advertisement is not doing anything and there is nothing shown on the TV set. The woman and the TV set are placed together in a way that is rather awkward to modern eyes.

The relationship in this ad between the TV set and the woman is different from the one between person and television in the American TV-set ads of the 1950s. According to Lynn Spigel, such ads often emphasized the interaction between the content on TV and the audience. In the *Kaige* TV-set ads, the reflection of light on the screen and the closed curtain tell us it is night, after working hours. Although the TV set is not turned on, the inviting smile on the woman's face shows us that she is waiting for the moment when she can begin to relax after a tough day. The woman's make-up, long hair, and clothing, combined with the TV set next to her, symbolize a young and modern lifestyle.

Mulvey (1992) and Berger (1972) famously argue that in patriarchal culture, images in the media replicate the male gaze. In other words, women are invited to identify with the images, and men are invited to gaze at them lustfully. According to this theory, the woman and the TV set in this American advertisement would both constitute symbols of what a man could own in modern life: a beautiful stylish wife and a shining TV set viewed together become an attractive combination. If this is the intention of the advertisement, we can easily see how quickly Chinese ideology changed in the 1980s. What would happen if we switched the woman sitting next to the TV set with a man? Would it also work well to promote the TV set? How would the man act if he were sitting next to a TV set?

These questions can perhaps be answered through an examination of additional TV-set images. Of the twelve volumes of *Popular TV* issued in the 1980s in China, only four had covers showing men: a monk from a TV drama, an extremely young and feminine-looking male singer, a man part of a western-looking couple, and two actors together with an actress at an awards ceremony. The remaining covers featured stylish and modern-looking female actresses.

Their clothes are far more modern than the clothing people usually wore at that time. The men who appear on the covers are all special figures. They are either desexualized men or cultural outsiders (see Figure 5.3 in Photo Spread).

Many advertisements are similar to one for Bai Hua (Hundreds of Flowers) TV. The advertisement selects three female figures from different TV dramas to accompany a TV set, which is located in the bottom-right corner of the advertisement (see Figure 5.2 in Photo Spread). The images of the women are seen as more important than the image of the TV set.

That only four men are depicted on the cover of *Popular TV* and that female images play a dominant role in the advertisements and magazines echo Mulvey and Berger's argument that media images are selected on the basis of the male gaze. It is not surprising to find this phenomenon in Chinese media. But the practice took on certain unique characteristics in its travel across space and time to China. Comparing to U.S. media, the Chinese media were clearly not comfortable displaying a masculine man by himself in TV advertisements or on cover of *Popular TV*.

However, the kinds of women shown in these advertisements are less commonly seen in the more serious magazines and literature, such as *Banyue Tan* (*Bi-Monthly Discussion*, an inside magazine of Chinese Communist Party). The difference between the women in *Popular TV* and those in *Bi-Monthly Discussion* is that the former occupy the role of either "being looked at" or being entertaining, while the latter express authority and decency. Although the socialist government stipulated that men and women should be completely equal in all ways, it was very difficult for people to operate according to this ideology in daily life. Modernization is not hard to implement if it is only a matter of technology, personal appearance, or lifestyle. What makes it complicated is the clash between an unspoken cultural sensibility and a new cultural environment. This contradictory relationship was apparent in China in the 1980s. Western ideology was quickly being imported but was clashing with traditional culture. The media still did not feel comfortable presenting men as objects for viewing pleasure in 1980s China.

It is hard to find ancient Chinese paintings of men who were not of high social status. There are paintings of emperors, generals, saints, and gods, but paintings of men drawn simply for the purpose of visual pleasure are rare. How might a man's face be portrayed without any specific political intention? There are some pictures of male actors and scholars in magazines, but most of them just accompany a biography, which is the central focus. This practice was still common in the visual culture of China in the 1980s, during the early era of modernization. Governmental declarations can change a situation officially but the ideology embedded in a culture changes at a much slower pace.

During the 1980s, the images of women in TV advertisements and on the covers of various entertainment magazines were mostly well polished. The images of women advanced in the sense of being stylish in a modern way. The Chinese phrase to describe them is "Shi Mao" (a more colloquial way of saying "modern," "stylish," or "trendy"). The women's make-up, eyebrows, and clothing are far removed from what most common people actually wore at the time. However, these images were intended to be models for the readers of the advertisements and magazines. Consequently, the traditional idealized portraits of women and men who are supposed to be respected and emulated for their morality and spirit and, fittingly, lack of concern for their own appearance, conflicted with the emphasis on female appearance and material products in popular magazines and commercials during the 1980s. Television as both a commercial product and carrier of dominant ideas is portrayed in such a complex way. While television programs served to educate people and remind them of the beauty of morality and spiritual power, television itself was promoted in advertisements that were highly conscious of the ideology of consumer culture and the "flaws" of both capitalism and feudalism—men should be looked at, and money could make women more beautiful to be looked at.

"Ignoring" Women: Masculinity and Femininity in *The Bell Tower and the Drum Tower* and *The Policemen in Plain Clothes*

Women returning to love and marriage, men returning to nation and people: these behaviors suggest that China was returning to the old days but the trend also contained an element of something more modern and westernized. Marriage and love are two of the most basic topics with which human beings have always struggled. In the 1980s, Chinese women were awakening from the mentality of collectivism in which they were supposed to think about their lives symbolically in terms of the larger cultural context, the nation, and the society. In the early twentieth century, male writers and directors were still thinking about women according to the traditional roles of man and woman in society. Marriage and love relationships are still two of the most common subjects to look at when evaluating women's morality and their attitudes toward their lives.

A clear trend appeared at the end of the 1980s: creators of TV dramas made more of an effort to have their stories reflect the lives of common people. Humanity is one of the main themes of Chinese intellectual discourse. This appeal to humanity had a very strong appeal after the Cultural Revolution, in which the extreme emphasis on "revolution" and "rebellion" caused people to lose their sense of humanity and think, in a purely symbolic way, in terms of class division.

Ussher (1997) has argued that the media provide scripts for gender identity, especially on how to be a woman. Her argument addresses the highly developed consumer culture of the West. However, the notion of "transcripts" is very applicable to media in general. The transcripts that media provide are never all-encompassing, despite the efforts of minority groups, such as feminists, lesbians, gay men, and others, to make invisible groups visible in recent decades in the United States. Chinese television, although it continued to be censored, became much more refreshing to audiences when it called for humanity and encouraged excitement over Economic Reform. Domestic television dramas could not compete in quantity with imported dramas, but the influence of the few that did exist should not be underestimated. As mentioned earlier, *The Snow City* expressed the idealized femininity and masculinity while naturalizing a new ideal of womanhood. The other two popular TV dramas of the period that had information to convey about the depiction of gender roles were *Zhong Gu Lou* (The Bell Tower and the Drum Tower) and *Bianyi Jingcha* (The Policemen in Plain Clothes).

The Bell Tower and the Drum Tower is based on a novel by Liu Xinwu and depicts the lives of common people in Beijing. The novel contains many characters, most of whom were kept in the TV adaptation. The TV drama kept the original flavor of the book but edited it to finish the story in eight chapters. The tone of the TV show is very relaxed. It was not as serious as other programs on television at the time and did not portray the intensive struggle of life in the city as did *The Snow City*. *The Snow City* focuses on the late 1970s and the early 1980s, when everyone in China was facing an extremely difficult transformation from the past.

The Bell Tower and the Drum Tower reflects China a few years after the Economic Reform, when modernization was moving ahead rapidly. People's formerly collective-based lives began to differentiate from one another. *The Bell Tower and the Drum Tower* attempted to represent people from as many different backgrounds as possible. The characters included retired elders, intellectuals, workers, officials, doctors, and unemployed people and characters of a variety of ages, occupations, and family backgrounds. The central narration focused on a few different families living within the same neighborhood. Each family had its own problems and joys. The story has several different plot threads. Gender seems to be merged to the other threads. There are many female characters. Mu Ying is a doctor who falls in love with an official living next door; Xiu Zao is a young college student who has a crush on a young intellectual in the neighborhood; Zhan Liying is a married middle-aged woman who is learning to be a matchmaker and whose husband is working in a different city. Each of these women represents a certain social group of women and each of them has a distinct character.

Mu Ying is twice divorced, which was very rare in China at the time (as it still is today). She is a revolutionary woman in a new era. She marries out of respect, admiration, and therefore love for a man, and she gets divorced when she realizes she no longer loves her husband. The TV drama does not judge this woman but represents her from an objective perspective. She is like many other women in several ways but her distinct experience with marriage highlights the gender issues important in 1980s China. Mu Ying's insistence on finding true love causes her to be cold in dealing with her children. She leaves with each ex-husband the child she had with him and completely cuts off her connection with the children. She is always looking to find her ideal man, someone with whom she can maintain a refreshing spiritual relationship. Her attitude toward her marriage may seem to represent a change from those of the traditional woman. However, in her strong persistence, it is easy to see the remnant of a historical ideal of love, which we see in *The Snow City*. She does not have to prove herself strong through a physical competition with men; rather, she wants to live by following her passion for love.

Mu Ying's passion for love seems like a rebellion or a challenge to women's traditional role and morality; however, analysis of her choice of men reveals that she is stuck within patriarchal thinking. Her first husband became a national hero after being disabled in the war; she fell in love with him after reading his heroic story. This marriage, unexpectedly, gave Mu Ying a chance to leave the countryside and go to the city. Although Mu Ying loved this man, when she went to college courtesy of the special arrangement the government made for the wives of heroes, she became infatuated with her college's young student president. She was touched by his speeches and ability to lead, and so divorced her first husband and married the student. However, soon after, Mu Ying becomes bored with her second husband. She begins to find him ordinary. Then she discovers that she has a passion for an older official, who was not treated fairly by the government during the Cultural Revolution. The older official's rich spirituality appeals so much to Mu Ying that she tells her husband and goes to tell the official what is on her mind. She is honest to her love, the only thing that she thinks important in her life. She is very elegant and beautiful and displays great social manner, peace of mind, and a sense of independence. However, every time she selects a man, he becomes her hero and his social background comes to define her own values. In every case, she thinks she is being very faithful to her personal passion, but each passion is clearly constructed by societal criteria.

Considered from either a traditional view or a modern view, Mu Ying's pursuit of honesty in love and freedom in marriage never manages to be as free as it seems. There is never a struggle between her and her husband when Mu

Ying decides to change her life. She is fortunate that each husband is in turn willing to set her free, although he still loves her. Mu Ying's attitude toward her children and other people's gossip about her challenges the prevailing opinion about what constitutes a good woman. Love determined by the social situation is more important to her than anything in her life, and her love constructs the different stages of her life.

Mu Ying's situation is an extreme case in *The Bell Tower and the Drum Tower*; however, her preoccupations, men, love, and marriage, are also dominant in the portraits of the other women characters, although the details differ. The women, no matter their age or educational background, are all preoccupied with men. The life of each is shaped by a man who they either are with or want to be with. Zhan Liying's enthusiastic performance of the role of matchmaker resulted, for instance, from her long-distance relationship with her husband. "She needed to be a matchmaker to make herself feel satisfied," as the TV drama narrated. In general, *The Bell Tower and the Drum Tower*'s emphasis on women's lives suggests a new consciousness of the role of women in society. Although both men and women worked and had equal opportunities, at a deep cultural level, people still thought women were most concerned with their emotions and personal affairs.

This consciousness is also reflected in *The Policemen in Plain Clothes*, a show about a police squad with only one woman member, Yan Jun. Although she works among men in a job traditionally reserved for men, by the end of the drama she succeeds in transferring, with her parents' help, to a more traditionally female job, in the Department of Train Communications. Yan Jun is called "Jia Xiaozi" (a tomboyish girl), but neither her appearance nor her character reflects this name; she is called this only because she is a woman working at a dangerous law enforcement job. *The Policemen in Plain Clothes* seems conflicted over how to present the character of the job and the character of a young woman as a person. Yan Jun's work performance in the department is related to her gender. She is patient in interrogations and can be a good actress when she is undercover. But what distinguishes her most from her coworkers is her unselfish love for the hero of the drama, a young police officer who already has a girlfriend and who cares more about his job and the nation than about his relationship.

It is clear that there were more diversified representations of people, especially women, in the TV dramas in the 1980s in China than in pre-Economic Reform China. However, although characters' appearances, backgrounds, experiences, and jobs differed in many ways, the core values of people's lives were still the same. The TV dramas were still presenting family, love, and marriage as women's main concern. This is not to say that there were no career-driven women in the TV dramas; Zhantai Zhizhu, a Peking Opera actress in *The Bell Tower and the Drum Tower*, was very devoted to her career. She was respected and envied because of her "perfect" life as a wife, a mother, a daughter-in-law, and an actress. But her life was still mainly devoted to her relationship with her husband who had a hard time appreciating her performance and was always jealous of her relationship with a particular actor. In contrast to the TV dramas' varying representations of women, the shows always portrayed the "good man" as unselfish, patriotic, knowledgeable, and helpful. These men cared more about others and the future of the country than about their personal or family concerns. In *The Bell Tower and the Drum Tower,* Xun Lei, a young intellectual heartthrob, seems more concerned with his studies, job, and country than with girls. The TV show's narrator does not comment about his attitudes toward love even when he is shown with his girlfriend or when he is made aware of a neighbor who has a crush on him. His ignorance of and indifference toward love seem to be why women love him. On the other hand, he often makes clear declarations of his love for his country, especially when he returns to China after his two years of study in England. A new hero at a new historical moment would serve the country with his knowledge and heart. A similar attitude is displayed by the old official to whom Mu Ying is attracted. The men in *The Snow City* similarly are most concerned with brotherhood, the morality of being a man, other people's happiness, and their comrades' future. These heroes' ignorance of women well exemplifies Antony Easthope's (1992) argument that masculinity is given by all patriarchal cultures:

> Masculinity aims to be one substance all the way through. In order to do this it must control what threatens it both from within and without. Within, femininity and male homosexual desire must be denied; without, women and the feminine must be subordinated and held in place. (p. 166)

Because these TV dramas present love and relationship as women's concerns, men, as the heroes, have more important concerns. They are given more important assignments to protect their masculinity from "within" and "without." Thus, the women who love their men secretly, adoring them all the more because of the men's indifference to love, are quite difficult to understand. No matter how virtuous or brave, the women are always cast in a secondary role as observers or encouragers of men's lives. And the women who want to interrupt or dominate over men's lives, such as one young bride in *The Bell Tower and the Drum Tower*, are portrayed as fetishist (because she asks for many things from her in-laws before she agrees to marry her fiancé), immoral, and bothersome.

Generally, Chinese TV in the 1980s presented men and women as two different types of creatures, a distinction which differs from the "iron girl" and the "iron man" propagated in the 1950s and the 1960s as icons for the nation. The word "iron" actually asexualized people, denying any essential differences that might possibly limit women's contribution to the nation. In the TV dramas of the 1980s, women were no longer iron-like; they were feminized either through appearance or personal traits, and men were accordingly remasculinized. All the important officials were males. The respectable men in the stories are the ones who have a vision beyond their daily concerns. They are concerned with their relationship with an imagined, bigger, otherness, such as the safety and happiness of common people, the development of other countries, and the well-being of people in general. Chinese masculinity has always been connected with the nation and the future of the nation. Masculinity in the 1980s focused on similar issues, while also acknowledging the importance of a man's knowledge and his ability to communicate with people who need his help. The media in the 1980s allowed some male characters to be sensitive and cry from time to time, but the heroes are always peaceful, objective, confident, persistent in their beliefs, and most important, knowledgeable or smart enough to continue their education. Their respect and admiration for knowledge are emphasized by the writers of the stories and the producers of the TV dramas, who are themselves male intellectuals. This respect for intellectuals reflects a turning back to the values of traditional elitist culture.

Kam Louie (2002) has argued that there are two types of masculinity in traditional Chinese literature. *Wen* (cultural attainment) masculinity had been considered superior to *Wu* (martial valor) masculinity before the People's Republic of China. Whereas men with *Wu* masculinity were able to resist feminine lure, *Wen* masculinity reinforced or represented romance with women. This dyad masculinity, as suggested by Louie, has evolved historically. The ideal of socialist masculinity is insensitive to women's affection in general. In post-Mao China, ignorance or indifference to women's affections is still an important characteristic of popular heroes. In the two TV dramas *The Bell Tower and the Drum Tower* and *The Policemen in Plain Clothes*, the heroes are intellectual and good-looking. However, they do not devote much time to love, relationship, or family. Instead, they care most about serving the country and doing their own work well.

Post-Mao masculinity is a curious combination of *Wen* masculinity and an indifference toward women, which was a characteristic of *Wu* masculinity. Different from femininity, which was given double meanings, masculinity in the 1980s was a reconstruction combining elements of each prototype from the traditional culture, resulting in a new gender norm appropriate to the modernization of China. The discourse of masculinity portrayed in 1980s Chinese media still involved loyalty and patriotism but added a concern with knowledge and trivialization of private life. The individuality spurred after the Cultural Revolution had to be expressed in service of the public that men could play. The dominance of male writers and producers in Chinese domestic TV-show production could not help expressing the narcissism of a remasculinized ideal toward life. It can be seen from the various trivial things that the stories portray women, especially annoying women, as being concerned with, such as matchmaking and betrothal gifts. Women characters in the TV dramas are excessively rude, selfish, snobbish, and histrionic, among which the stepmother of Xu Shufang in *The Snow City*, the older daughter of the mayor in *The Policemen in Plain Clothes*, and the annoying bride and her aunt of *The Bell Tower and the Drum Tower* are all examples. The portraits the TV dramas paint of these unlikable female figures are much more vividly detailed than are those of the flawed men. The unsympathetic male characters err by being lazy or criminal. In other words, it was often up to a boss or the political system to make a judgment on men; however, that which constituted goodness or badness in women was much more fluid, which is why the topic invited such thorough evaluation by TV dramas. Issues of women's morality often lay beyond the clear-cut criterion of authority; in fact, even when a woman committed a crime, it was often for morally ambiguous reasons, such as the ambition for a good future, or the result of being misused by a criminal man, or based on a misunderstanding of "liberation." Lu Tonglin (1993) has argued that 1980s Chinese literature is heavily misogynistic, and she illustrates the relationship between the misogynistic discourse and the social context.

> To a large extent, the misogynistic discourse prevailing in contemporary Chinese literature can be traced to the salvation theory of the socialist revolution of which socialist realist literature is one of the best expressions. This discourse reinforces the underlying gender hierarchy existing in the socialist salvation theory. The Communist Party, as a collective savior in socialist realism, usually proves its ideological and political superiority by the salvation of silenced and oppressed women. In the same vein, an individual hero in contemporary fiction very often reaffirms his masculinity or individuality in opposition to women's inferiority and subordination. In other words, in order to regain his individuality repressed by the communist domination, the male individual of contemporary fiction needs to use the female body as a scapegoat for communist ideology. In both cases, women, situated at the bottom of the hierarchy, function mainly as symbols for their superiors, be they communists or individualists. (Lu, 1993, p. 3)

It is not an exaggeration to say that depictions of women in general were of their personal lives, while depictions of men in general were of the situation and the policy of the nation. The TV dramas portray women and men living in a reality created by the dominant ideology and the governmental policy, but it is always in the men's stories that we see the situation of the nation and society.

The shows' insistence on the differences between men and women was a necessary way to remasculinize and refeminize. For example, refeminization is reflected when the female police officer, Yan Jun, in *The Policemen in Plain Clothes*, buys flowers and brings them to the office. (Interestingly, on an episode of the American TV detective show *Hunter*, a woman detective does the same thing.) Although Yan Jun was called "Jia Xiao Zi" (the tomboyish girl), she is very caring, patient, and romantic, too. She can cry.

The media legitimized different masculine and feminine traits, as was required by modernization and economic reform. Once people were allowed to return to the cities and focus on improving their lives within a family unit, instead of within a corps or other bigger social team, the gender roles in each family began to return to their traditional pattern. However, this was not a complete return. Masculine men were required to continue to think about their relationship with the bigger world, whereas women were allowed to care about some tedious, seemingly trivial things. Although Western feminism might find this to be a decrease in women's status, in 1980s China, turning away from the previous three decades empowered women to regain their separate identity. Having men express political thoughts while women chat and even gossip about such matters as appearance, relationships, and food reasserted femininity after a period of celebrating iron girls and privileging of military power.

Importing Masculinity (and Femininity): American TV dramas

Importing foreign TV shows was a necessity for Chinese TV stations in the early 1980s. The tremendous increase in the number of people owning TV sets and the establishment of local TV stations required much programming to fill the available time slots. There were not enough domestically produced TV programs to fill these hours. At the same time, audiences found the domestic TV dramas not sufficiently entertaining. Therefore, both Chinese Central TV and local TV stations began to import TV shows. American TV shows predominated because of the affordable rates extended to Chinese TV stations (Hong, 1998). *The Man from Atlantis (*1977/1980*)*, *Garrison's Gorillas* (1967/1980), and *Hunter* (1983), played throughout China, were the three most popular American TV dramas of the period. Interestingly, all three dramas feature extremely masculine protagonists. In addition to these, many Hong Kong and Japanese TV shows were imported into China in the 1980s.

The relationship between men and women does not show a revolutionary change in the imported media in the 1980s. In *Hunter*, the protagonist detective's partner is a brave and smart woman. Many similar women characters appear in *Garrison's Gorillas*. A woman doctor saves the man in *The Man from Atlantis*. Many independent and smart women also appear in Hong Kong TV shows and movies and Japanese TV shows. However, despite the courage and talent of these women in the imported TV shows, they are still mainly portrayed as assistants to the heroes. Thus, femininity is basically a supplementary element in the portrait of masculinity. Some of the women characters, especially those in Japanese TV shows, are victimized. Women serving as assistants to men in these TV shows is mainly a result of the stories' historical context. Thus, an examination of the presentation of masculinity in these TV shows is central to understanding the following discussion.

Masculinity in 1980s Western TV programs differed from the masculinity portrayed by Chinese intellectuals in novels or TV shows. Geng Song (2004) proposes an explanation for the distinction between Western and Chinese masculinity. He believes that "the definition and representation of masculinity in pre-modern China show the unique characteristics of male subjectivity in Chinese culture" (p. 171).

> In Western culture, masculinity is defined firstly as the power and ability to conquer nature, which is symbolized by woman. This is a reflection of the Self/Other dichotomy. Women have been treated as the other and as a symbol of nature, as the site of male conflict and conquest. Masculinity is therefore defined mainly as an individual subjectivity and by man's relations with woman/other/nature.
>
> In the Chinese cultural tradition, however, masculinity or gender refers mainly to a position in the political structure. The Confucian prescriptions for a real man include filial piety to his parents, obligations to family (among them the most important one is the ability to carry on the family line), and above all, loyalty and contribution to the sovereign and state.
>
> In other words, masculinity is not defined by a man's relations with his woman, but by his relations with the political mechanism. For Chinese men, manhood is the ability to honor their family name and achieve fame in serving the state. This is to be accomplished in the public world. It is therefore not difficult to understand why sexuality was absent in the construction of masculinity in the heroic discourse. (p. 171)

The ideal of Chinese masculinity was a man who was intellectual, patriotic, knowledgeable, diligent, honest, serious, caring, selfless, philosophical, and marked by strong self-control. Masculine men, it was generally believed, should care more about nation and people in general than about their girlfriends or wives. They were to work hard for the happiness of people in the world at large, rather than their own family members. The appearance of this type of idealized men is not uniform but is most defined by an intellectual view, even when the men are doing physical work in TV dramas. The Western masculinity ideal privileged a strong physical figure, humor, defiance of authority, play-boyish charm, and unique personality. The men in the imported Western TV dramas are mysterious, like the man from the ocean in *The Man from Atlantis*, a figure of science fiction who is cool, calm, and handsome. *Garrison's Gorillas* features four men who are all from different backgrounds. Each of them is somewhat of a deviant, because he has committed a crime according to his individual "talent." In *Hunter*, the title character is from a lower-class family and is not afraid to offend the authority of the high-rank police officers.

Chris Blazina (2003) has argued that in Western culture, masculinity has always been evolving, and so examining the cultural myth of masculinity throughout history is helpful in "understanding the making of modern middle-class masculinity" (p. 46). He believes that modern Western middle-class masculinity inherited the ethic of medieval chivalry, which "attempted to blend together aspects" of the "differing definitions of prototypical masculinity." The differing aspects, according to Blazina, are the valor of warriors and the refinement of aristocrats. He says, "This paradigmatic conflict was not well resolved by either the ruling aristocracy or the warrior class before the advent of the middle class" (ibid). Blazina further points out:

> In essence, the middle-class solution to the two competing masculinities is the same one the mythmakers of the Middle Ages attempted. Masculinity is dichotomized as reflecting either refinement or valor. These, in turn, are attached to social standing whereby refinement is indicative of a higher standing and valor alone may be suspect of a lower standing. It is only the rare individual in myth or reality who can straddle both worlds successfully . . .
>
> The men of the new, burgeoning middle class found themselves struggling with issues of nobility and valor. The middle-class man was now attempting to be a gentleman through action, not pedigree. (ibid.)

Blazina's examination illustrates why masculinity as presented in 1980s American TV shows, such as the ones we consider here, has such conflicting characteristics. The three TV dramas that were very popular in China in the 1980s always based the attractiveness of the male protagonists on their defiance of authority, their challenges to the rules meant to regulate their behavior, and their subsequent power to "rule" others. Nothing seems more important to them than values such as fairness and freedom. They love innocent people and protect virtuous women and would mercilessly kill a criminal for the purpose of protecting other "good" people. As Blazina argues, these characteristics reflect chivalry. These masculine figures always deny their fear and undermine authority, but they are portrayed as standing on the side of ultimate justice; their outstanding, unique personalities are indifferent to the official protocol of their jobs. This Western masculine figure, who was rebellious and seldom engaged in a serious conversation with friends, yet very serious about justice, was the model often shown on Chinese television during the 1980s.

The individualist modern knights were beyond the imagination of Chinese media at that moment; however, they were everywhere in the Hong Kong martial arts chivalry TV shows and movies imported into China in the 1980s. Hong Kong media, like the Western media, attempted to construct a modern middle-class masculinity. However, because of Hong Kong's heritage of traditional Chinese culture and its need for the spiritual support of their colonizing culture, Hong Kong media often expressed a type of quasi-nationalism. This nationalism is not like the nationalism of mainland China, which had a specific and clear concept of "nation." The nationalism portrayed in Hong Kong media constituted more of a loyalty to the ethnicity and a sympathy with people in the lower economic classes within the Chinese colonies. Hong Kong media celebrated the dramatic and legendary success of "small" people and created stories in which lower-class youth got the chance to be heroes and help their fellows. This masculinity is a combination of individualist Western chivalry and traditional Chinese masculinity—it imagines the talent of heroes while valorizing relationships with others.

As Blazina (2003) has argued, modern Western middle-class masculinity was shaped by a change in the economic environment. With the onset of mass production and the division of labor, people gained more time for leisure and socializing. As consumer cultures develop, men who want to court women must pay attention to their own appearances. Masculinity is judged not only by behavior but also by personal physical appearance. Therefore, given their similar economic state, it is not hard to understand why the masculinity portrayed in Hong Kong media shares similarities with that of American media of the 1980s, even if the stories are different. However, the masculinity portrayed in the media of mainland China has none of the characteristics celebrated in Hong Kong or American media.

As noted, foreign dramas were imported into China out of necessity—the increasing numbers of people owning TV sets and local TV stations being founded and the lack of domestic TV programs. At the time, as Hong Junhao (1998) argues, the "affordableness" beat the "acceptableness" of foreign media and brought many "artistically inferior foreign shows" into China (p. 64). However, these "inferior foreign shows" could still expose the kernel of foreign cultural values, maybe even more than could the classic shows. Historical background and cultural context influence masculine ideals. Masculinity is always evolving. Although it is hard to know the exact extent to which Chinese people have accepted the representation of masculinity in the imported media, the popularization of "Mark Sunglasses" in China after the broadcasting of *The Man from Atlantis* (Guo, 1997, p. 25), and the numbers of youth wearing a white scarf after watching the hero of *Shanghai Bund*, probably the most popular Hong Kong TV show in the 1980s China, to some degree indicate the relationship between imported foreign media and local Chinese culture. At least, it shows a beginning of a trend wherein masculinity is expressed through the consumption of inexpensive, trendy items.

Conclusion

Political mandates are powerful. However, no matter how much control government has over constructing people's beliefs and behaviors, if the essential kernel of a culture's ideology about gender identity and gender relationship is not changed, the norms will persist.

Even in socialist China, where there was a clear mandate that men and women be equal, traditional cultural norms of gender were never changed, though they were shaken a bit. Although men and women officially are allowed to work in any occupation, their equality is still superficial and undermined by a problematic patriarchal ideology. The importation of American media into China, at a time when issues of gender identity were still unresolved, added greater complexity to an already confusing picture. The extremely patriarchal stories and images and the modern middle-class masculinity of American culture reinforced the 1980s Chinese media's return to traditional gender norms.

EPILOGUE
CONTINUITY, IMPORTATION, CREATION AND THE CHINESE TV INDUSTRY

As I have argued, technology studies, globalization studies, and gender studies all evolve according to historical and academic developments. In this book, I have had to limit my study to a manageable scope, but I do not think cutting off the 1980s from the 1990s, especially the early 1990s, is optimal. Society has its inertial property; so does media development. Therefore, the periodization in this book includes both the late 1970s and the early 1990s. In fact, television had an even more complicated experience in China in the 1990s than in the 1980s.

New TV Sets and New Advertisements

TV-set advertisements in the 1990s were different than those in the 1980s. Many well-known TV-set brands disappeared in the 1990s, among which Golden Star, Peony, West Lake, Panda, Kai Ge (Victory Song), and Lily have been mentioned in this book. Advertisements for TV sets infrequently appeared in the print media in the 1990s. For example, *Popular Cinema* published many TV-set advertisements in the 1980s and also advertised factory giveaways of sets to stations that produced enough high-quality television programs. However, during the 1990s, *Popular Cinema* rarely printed any images or news about television. Even the critiques of television programs that often appeared in *Popular Cinema* in the 1980s disappeared in the 1990s. The relationship between television and film was no longer a rivalry. The television industry had grown into a media giant since the late 1980s, thanks to the sacrifices of pioneers in early TV-set factories and to the celebration of new electronics companies by the developing consumer culture. Most people did not doubt the necessity of owning a TV set in 1990s China. The presence of TV sets in people's homes was already naturalized.

In 1989, Chang Hong TV Sets Group, which had been a Chinese defense-industry factory from 1958 to 1965, began promoting TV sets with the slogan, "Chinese Chang Hong" and initiated the first price war between Chinese color TV sets and imported color TV sets. The affordable price of Chang Hong for many Chinese households allowed the domestic TV-set producer to win the war. In 1996, Chang Hong, which by then had evolved into Chang Hong Electronics Group, unrolled a new slogan: "The Sun is redder, Chang Hong is newer." Chang Hong since has become one of the largest electronics manufacturing groups in China. The slogan used by Chang Hong and the red color that often appears in Chang Hong's advertisements echo two general themes that have been promoted by TV-set producers in China since the early 1980s: nationalism and nature. The Chinese character "Hong" in Chang Hong means rainbow, and the sound of "Hong" is the same as that of "Red" in Chinese. Moreover, "Sun is redder, Chang Hong is newer" announces a more advanced modern electronics group in a stronger China.

Chang Hong had competitors from both domestic and international TV-set manufacturers, among which Kangjia (Konka), Chuangwei (Skywatch), TCL, Panasonic, Sony, and Toshiba are well known in China. Most of these companies advertise their products on television. These companies create their own slogans and emphasize the uniqueness of their own products. It is clear that TV sets have become a capital commodity, rather than solely a technological product used to symbolize a modernizing culture, as they were represented in the Chinese media during the 1980s. During the 1990s, advertisements tried to persuade people to choose among different TV-set brands rather than trying to convince people to buy one at all. Consequently, advertisers needed to emphasize the uniqueness of each brand, which resulted in the TV-set advertisements being much more interesting and attractive than those in the 1980s. The promotion of TV sets still included the older themes of nationalism and nature.

Yearning, Chinese TV Drama Productions, and Chinese TV Studies

In 1990, a fifty-episode television series, *Ke Wang* (Yearning), coproduced by Beijing TV Stations and Beijing TV Art Center, gained a big following in China. Scholars have conducted detailed analyses of this series and its commentary about the Cultural Revolution as well as about common people's yearning for a new life at the transitional moment between the late 1980s and early 1990s. What scholars have not examined is what the producers of this series learned from the television programs produced in 1980s China; nor have they examined how it was influenced by imported TV dramas.

Many TV series were produced in China in the 1980s, but none gained the viewership that *Yearning* did. *Yearning* was the longest-running TV series when it was produced in 1990, and according to Zha Jianying's (1995) investigation, it was the first Chinese TV series to be based on a screenplay rather than on a well-known literary work, as were the 1980s TV dramas. Zha interviewed *Yearning*'s writers and producers who, because of the show's success, are recognized as the pioneers of Chinese soap operas. These writers and producers recalled how assiduously they worked to finish the scripts and how they embedded their yearning for a better life into the dialogue. These producers and script writers since have become important figures in the production of TV dramas in China. One of the directors of *Yearning*, Zheng Xiaolong, has since directed many other TV shows.

Although *Yearning*'s success led to it being the topic of many academic studies, these studies paid insufficient attention to certain aspects of the show, including gender issues (no women are mentioned when people talk about the production credits); how *Yearning* incorporated unique aspects of Chinese history and culture within an imported TV genre, the indoor soap opera; and how *Yearning* created a successful model for TV dramas in China. The heroine of *Yearning* brings to mind the women figures portrayed on Chinese television programs during the 1980s. In *Yearning*, the personality of the beautiful, kind Liu Huifang is in some ways similar to that of Xu Shufang in *The Snow City*. However, scholars have not compared them by tracing the history and engaging in a deeper analysis of the representation of gender on Chinese television. In addition, studies of *Yearning* are also typical of TV drama studies in failing to consider the TV production industry and its relationship with Chinese TV stations and the global market, in other words, how the production, distribution, and broadcasting of TV dramas have been managed in China since the late 1980s when many TV art centers/TV studios were established. *Yearning* is an inspiration and encouragement to the TV art centers that are independent of TV stations, but at the same time, it is still a co-production of a station and an art center. A study of the production and distribution of *Yearning* would be very helpful for a clear understanding of the history of Chinese TV industry. Paradoxically, while scholars such as Hong Junhao (1998) have examined the distribution and broadcasting of imported TV dramas in China since the 1980s, there are not enough academic studies of Chinese domestic TV dramas. Especially because many actors who started their career in the 1980s TV dramas have created their own media production companies, it seems that scholars would be more curious about the connection between TV production and distribution in the Chinese TV industry.

As I argued in the introduction, studies concerning Chinese television since the 1990s have proliferated. In *Television in Post-reform China*, Zhu Ying has examined how television dramas have developed as "art, political discourse, and transnational capital" (2008, p. 1). Zhu's study discusses the general relationship between television and dominant ideology: the latter determines what can be aired on television, and television reflects and reinforces the dominant ideology; however, the global television market complicates the TV content aired in China. Television dramas imported by China often contain novel content or content that conflicts with the dominant ideology. Moreover, to compete with TV shows produced in other parts of the world, Chinese television producers also needed to learn from others, while maintaining the uniqueness of Chinese culture. Consequently, for a clear understanding of globalization and television productions, it is very important to examine the relationship between Chinese domestic television show production and the international television industry. The similarities between Chinese TV dramas targeting young audiences and those produced by Japanese and Korean TV industries for the same market also illustrate the necessity of conducting a study of the different industries. In the future my studies will illuminate the extent to which Chinese television productions have been purposely based on shows produced in other countries, and the extent to which they have struggled to establish their own uniqueness. Moreover, it will provide a clear picture of the relationship between TV productions and globalization, glocalization, regionalization, and hybridity.

Restless TV, Restless Media and Busy Audiences

Thanks to media technology innovations, especially the development of the Internet and the increasing competition among different media, audiences and users are immersed in a restless modern media culture. Although the gap between remote villages and modern cities has been increasing since the beginning of the Economic Reform in the 1980s, people in most parts of China are able to spend much of their leisure time on media. Chinese Central TV has sixteen uncapped cable channels, twelve HD cable channels, and twenty-eight Internet TV channels (CCTV.com). CCTV covers the whole nation as well as other countries and areas through satellite broadcasting. People in even the most remote parts of China can watch CCTV's programs. In the areas of China with more advanced technology, as in the developed nations, television has entered the stage dubbed "television after TV" (Spigel and Olsson, 2004). Media synergy, Internet TV, twenty-four-hour cable channels, and the ubiquity of TV sets all contribute to the now-common sense of restless TV and restless media. Consequently, we as audiences are restless, too. We can watch television on cell phones no matter where we are or what time it is, and we also can constantly surf the Internet. At the same time, we are living in a convergence culture, as Henry Jenkins (2006) calls it, in which we produce or reproduce videos or music by ourselves and share them with others through the Internet. Similarly, we can reproduce TV shows or upload clips of TV shows online to share them with others for free. This cultural environment makes media production, distribution, and usage much more complicated than they were in the 1980s. As the culture is restless, so are the scholars.

Last Words

I still remember when I was very young, and my mother would take me to a neighbor's home to watch television. The neighbors had saved their money and bought a 12-inch TV set in 1981 (my family could not afford one until 1985). The TV set was my biggest pleasure but my mother's biggest trouble. She hated to disturb my neighbor but could not stand my crying, either, so she tried her best to limit our visits. One night, after all three TV channels finished their broadcasting and the anchors had said, "Thank you, see you," I still did not want to leave and, while crying, I asked all the adults, "Why did they leave? Please let them continue . . ."

How many attempts do parents have to make to get their children to go to bed now? If there is a child like me, what can the parents do now when the TV is so restless? I dare not imagine.

BIBLIOGRAPHY

Allert, B. (1996). Languages of Visuality: Crossings between Science, Art, Politics, and Literature. Detroit: Wayne State University Press.

Altman, R. (1986). Television/Sound. In T. Modleski (Ed.), Studies in Entertainment: Critical Approaches to Mass Culture (pp. 39-54). Bloomington: Indiana University Press.

Appadurai, A. (1996). Modernity at Large: Cultural Dimensions of Globalization. Minneapolis: The University of Minnesota Press.

Baughman, J. (2006). The Republic of Mass Culture: Journalism, Filmmaking, and Broadcasting in America since 1941. 3rd ed. Baltimore: Johns Hopkins University Press.

Beck, U. (2000). What Is Globalization? Translated by Patrick Camiller. Malden, MA: Balckwell Publishing.

Beniger, J. (1986). The Control Revolution: Technological and Economic Origins of the Information Society. Cambridge, MA: Harvard University Press.

Berger, J. (1972). Ways of Seeing. London: British Broadcasting Corporation and Penguin Books.

Berman, M. (1988). All That is Solid Melts into Air: The Experience of Modernity. New York, NY: Viking Penguin.

Blazina, C. (2003). The Cultural Myth of Masculinity. Westport, CT: Praeger.

Bordo, S. (2004). Unbearable Weight: Feminism, Western Culture, and the Body. Berkeley: University of California Press.

Borgmann, A. (1984). Technology and the Character of Contemporary Life: A Philosophical Inquiry. Chicago: University of Chicago Press.

Brunsdon, C. (2003). The Feminist, the Housewife, and the Soap Opera. Oxford Television Studies. New York: Oxford University Press.

Callon, M. (1987). Society in the Making: The Study of Technology as a Tool for Sociological Analysis. In Thomas P. H., W. Bijker, and T. Pinch (Eds.). The Social Construction of Technological Systems: New Directions in the Sociology and History of Technology (pp. 83-103). Cambridge, MA: M.I.T. Press.

Carey, J. (1989). Communication as Culture: Essays on Media and Society. London: Unwin Hyman Inc.

Chai, C., and Chai, W. (1965). The Sacred books of Confucius and Other Confucian Classics. New Hyde Park, NY: University Books.

Chang, T. (2002). China's Window on the World TV News, Social Knowledge and International Spectacles. Cresskill: Hampton Press.

Chang, W. H. (1989). Mass Media in China: The History and the Future. Ames: Iowa State University Press.

Chinese Central TV Study Studio. (Ed.). (1999). The First person on Chinese Screen—Shen Li [Zhongguo Yingping Diyi Ren-Shen Li]. Beijing: Chinese Broadcasting and TV Publishing House.

Chinese State Council Study Studio. (Ed.). (2002). The Discussion by the Officials of Chinese State Council on China's Affiliation in WTO: Opportunity,

Challenge, Countermeasure [Zhongguo Jiaru WTO: Jiyu, Tiaozhan, Duice] Beijing: Chinese Yanshi Publishing House.
Chinese Pictorial [Zhongguo Huabao]. (1986). Beijing: Chinese Pictorial Publishing House, p. 5.
Chinese Yearbook of Broadcasting and Publications [Zhongguo Guangbo Chuban Nianjian]. (1986). Beijing: The Publication House of Chinese Yearbook, pp. 5-17.
Chu, G. C. (Ed.). (1978). Popular Media in China: Shaping New Cultural Patterns. Honolulu: Published for the East-West Center by the University Press of Hawaii.
Chun, W. H. K., and Keenan, T. (2006). New Media, Old Media: A History and Theory Reader. New York: Routledge.
Cui, S. (2003). Women through the Lens: Gender and Nation in a Century of Chinese Cinema. Honolulu: University of Hawaii Press.
Curtin, M. (2007). Playing to the World's Biggest Audience: The Globalization of Chinese Film and TV. Berkeley/Los Angeles/London: University of California Press.
D' Acci, J. (1994). Designing Women: The Case of Cagney and Lacey. Chapel Hill: UNC Press.
Donald, S., Keane, M., and Yin, H. (2002). Media in China: Consumption, Content and Crisis. London; New York, NY: RoutledgeCurzon.
Dong, M. Y. (2003). Republican Beijing: The City and Its Histories. Berkeley: University of California Press.
Douglas, S. (1994). Where the Girls Are: Growing up Female with the Mass Media. New York: Time Books.
______. (2004). Listening in: Radio and American Imagination. Minneapolis: The University of Minnesota Press.
Easthope, A. (1992). What a Man's Gotta do. New York: Routledge.
Esherick, J. (2000). Remaking the Chinese City: Modernity and National Identity, 1900-1950. Honolulu: University of Hawai'i Press.
Evans, R. (1994). Deng Xiaoping and the Making of Modern China. New York: Viking.
Evernden, N. (1992). The Social Creation of Nature. Baltimore and London: The Johns Hopkins University Press.
Fang, L. (2008, Sep. 28). Chenduo: Leading the Audiences to Travel. Xin Jing Bao (New Beijing). Retrieved Feb. 2, 2009, from www.gmw.cn/01wzb/2008-09/28/content_843392.htm.
Felski, R. (1995). The Gender of Modernity. Cambridge: Harvard University Press.
Feuer, J. (1992). "Reading Dynasty: Television and Reception Theory." In Jane Gaines (Ed.). Classical Hollywood Narrative: The Paradigm Wars (pp. 276-93). Durham, NC: Duke University Press.
Fiske, J., and Hartley, J. (2003). Reading Television. London and New York: Routledge.
Florida, R. (2005). Cities and the Creative Class. New York: Routledge.

Foucault, M. (1990). The History of Sexuality (Volume I): An Introduction. New York: Vintage Books.

French, D., and Michael, R. (Eds.). (2000). Television in Contemporary Asia. New Delhi: Sage Publications.

Gao, X., and Wu, Q. (2002). Er'shi Shiji Zhongguo Dianshiju Shilun [The Historical Studies of Chinese TV Plays in the 20th Century]. Beijing: Xueyuan Chubanshe.

Garnham, N. (2000). Emancipation, the Media, and Modernity: Arguments about the Media and Social Theory. New York: Oxford University Press.

Giddens, A. (1990). The Consequences of Modernity. Stanford, CA: Stanford University Press.

Giles, J. (2007). Gender and Consumption: Domestic Cultures and the Commercialization of Everyday Life. Burlington: Ashgate.

Goffman, E. (1979). Gender Advertisements. Edited by Harper colophon. New York: Harper and Row.

Gold, T. (1993). "Go with Your Feelings: Hong Kong and Taiwan Popular Culture in Greater China." China Quarterly, 136, 907-25.

Gunaratne, S. A. (Eds.) (2000). Handbook of the Media in Asia. New Delhi: Sage Publications.

Guo, Z. (1997). Zhongguo Dianshi Shi [The History of Chinese TV]. Beijing: Wenhua Yishu Chubanshe [Culture and Arts Publication].

Hartley, J. (2008). Television Truths. Oxford: Blackwell Publishing.

Havens, T. (2006). The Global Television Marketplace. London: BFI Publishing.

He, L. (2005). Zhongguo Dianshi Jilupian Shilun [A Study on the History of China's TV Documentary]. Beijing: Communication University of China Publishing House.

Hilmes, M. (2007). Only Connect: A Cultural History of Broadcasting in the United States. Belmont, CA: Thomson Wadsworth.

History of China's Television: Television Broadcasting. (2009, Mar. 1). Urban stations, CCTV.com. Retrieved June, 19, 2009 from www.cctv.com/tvguide/tvcomment/wtjj/dsrshss/10536.shtml

Hong, J. (1998). The Internationalization of Television in China: The Evolution of Ideology, Society, and Media since the Reform. Westport, Connecticut, and London: Praeger.

Hsü, I. C. (2000). The Rise of Modern China. New York: Oxford University Press.

Hu, Z. (2000). Zhongguo Dianshi Guannian Lun [The Critical Studies on the Notion of Chinese TV]. Beijing: Beijing Normal University Publishing House.

Huang, Y. and Yu, X. (2003). Broadcasting and Politics: Chinese Television in the Mao Era, 1958-1976. In Miller, T. (Ed.). Television: Critical Concepts in Media and Cultural Studies (pp. 354-368).

Hwang, K. (1999). "Filial Piety and Loyalty: Two Types of Social Identification in Confucianism." Asian Journal of Social Psychology, 2, 163-183.

Ikels, C. (2004). Filial Piety: Practice and Discourse in Contemporary East Asia. Stanford: Stanford University Press.

Innis, H. (1951). The Bias of Communication. Toronto: University of Toronto Press.
Iwabuchi, K. (2004). Feeling Asian Modernities: Transnational Consumption of Japanese TV Dramas. Hong Kong: Hong Kong University Press.
______. (2002). Recentering Globalization: Popular Culture and Japanese Transnationalism. Durham: Duke University Press.
Jenkins, H. (2006). Convergence Culture: Where Old and New Media Collide. New York and London: New York University Press.
Jones, A. F. (2001). Yellow Music: Media Culture and Colonial Modernity in the Chinese Jazz Age. Durham, NC: Duke University Press.
Kafka, F. (1971). The Complete Stories, edited by Nahun N. Glatzer, 235-249. New York: Schocken Books.
Kaika, M. (2005). City of Flows: Modernity, Nature, and the City. New York: Routledge.
Keane, M., Fung, A. Y. H., and Moran, A. (2007). New Television, Globalisation, and the East Asian Cultural Imagination. Hong Kong: Hong Kong University Press.
Kilborn, R. W., and Izod, J. (1997). An Introduction to Television Documentary: Confronting Reality. New York: Manchester University Press; distributed exclusively in the USA by St. Martin's Press.
Kim, A. S. (2002). East Asian and Globalization. Lanham, Md: Rowman and Littlefield Publishers.
Kittler, F. A., and Johnston, J. (1997). Literature, Media, Information Systems: Essays. Amsterdam: G+B Arts International.
Kittler, F. A. (1999). Gramophone, Film, Typewriter. Stanford: Stanford University Press.
Kraidy, M. M. (2005). Hybridity, or the Cultural Logic of Globalization. Philadelphia: Temple University Press.
Kynge, J., and Rahul, J. and (2000, Feb. 29). China on the Line. *Financial Times*, 22.
Larkin, B. (2004). Degraded Images, Distorted Sounds: Nigerian Video and the Infrastructure of Piracy. Public Culture 16 (2), 289-314.
Latham, K. (2007). Pop Culture China!: Media, Arts, and Lifestyle. Santa Barbara: ABC-CLIO.
Lee, C. (Ed.). (2003). Chinese Media, Global Contexts. London and New York: RoutledgeCurzon.
Lent, J. A. (1977). Third World Mass Media and Their Search for Modernity: The Case of Commonwealth Caribbean, 1717-1976. Lewisburg: Bucknell University Press.
Li, C. (2001). Korean Culture in China, Chinese Culture in Korea—The Proposal for Further Study on Korean and Chinese Culture. Modern Korea, winter.
Li, J. (1994). China's Media, Media's China. Boulder: Westview Press.
Li, Z. (1994). Women's Consciousness and Women's Writing. Trans. Z. Hong. In C. K. Gilmartin and G. Hershatter (Eds.). Engendering China: Women,

Culture and the State (pp. 299-317). Cambridge and London: Harvard University Press.

Liu, Y. (2003, Dec.16). The Transformation of Cultural Ecology of Chinese TV [Shi Lun Zhongguo Dianshi Wenhua Shengtai de Zhuanxing]. Retrieved July 5, 2009, from Academic.ChinaMedia.com. academic.mediachina.net/article.php?id=1511.

Lotz, A. (2006). Sex, Careers, and Mr. Right in Comedic Dramas: The "New" New women of Ally Mcbeal and Sex and the City. In Lotz, Amanda. (Ed.). Redesigning Women: Television after the Network Era (pp.88-117). Urbana and Chicago: University of Illinois Press.

Louie, K. (2002). Theorising Chinese Masculinity: Society and Gender in China. New York: Cambridge University Press.

Lu, S. H. (Ed.). (1997). Transnational Chinese Cinemas: Identity, Nationhood, Gender. Honolulu: University of Hawai'i Press.

______. (2000). Soap Opera in China: The Transnational Politics of Visuality, Sexuality, and Masculinity. Cinema Journal, 40 (1), 25-47.

Lu, T. (Ed.). (1993). Gender and Sexuality in Twentieth-century Chinese Literature and Society. Albany: State University of New York Press.

Lull, J. (1991). China Turned On: Television, Reform, and Resistance. New York: Routledge.

Lutz, C., and Collins, J. (2003). The Photograph as an Intersection of Gazes. In Wells, Liz. (Ed.). The Photography Reader (pp. 354-75). London and New York: Routledge.

Ma, B. and Guo, X. (1995). An Evaluation Study of Parent School in China. China National Institute of Educational Research. UNESCO Education Sector, Monograph, No. 2.

Marc, D., and Thompson, R. J. (2005). Television in the Antenna Age: A Concise History. Malden: Blackwell.

Marvin, C. (1988). When Old Technologies Were New: Thinking about Electric Communication in the Late Nineteenth Century. New York and Oxford: Oxford University Press.

Maynard, M., and Tian, Y. (2004). Between Global and Glocal: Content Analysis of the Chinese Web Sites of the 100 Top Global Brands." Public Relations Review 30, no. 3, 285-91.

McCracken, E. (1993). The Cover: Window to the Future Self. In McCracken, Ellen. (Ed.). Decoding Women's Magazines: from Mademoiselle to Ms. (pp. 13-37). New York: St. Martin's Press.

McKibben, B. (1995). The Age of Missing Information. New York: Plume.

McLuhan, E., and Frank, Z. (Eds.). (1995). Essential McLuhan. New York: Basic Books.

Mehta, N. (2008). Television in India: Satellites, Politics, and Cultural Change. New York: Routledge.

Meyer, M. (2000, Mar. 13). Generation Next. Newsweek, 48.

Miller, T., Govil, N., McMurria, J., Maxwell, R., and Wang, T. (Eds.). (2005). Global Hollywood 2. London: British Film Institute.

Moores, S. (2000). Media and Everyday Life in Modern Society. Edinburgh: Edinburgh University Press.
Morley, D. (2000). Home Territories: Media, Mobility, and Identity. New York: Routledge.
Morris, N., and Waisbord, S. (Eds.). (2001). Media and Globalization: Why the State Matters. Lanham/Boulder/New York/Oxford: Rowman and Littlefield Publishing, Inc.
Morris, N. (2002). The Myth of Unadulterated Culture Meets the Threat of Imported Media. Media, Culture and Society, 24, 278-89.
Moss, M. H. (2008). Toward the Visualization of History: The Past as Image. Lanham: Lexington Books.
Mulvey, L. (1992/2001). Visual Pleasure and Narrative Cinema. In Meenakshi, G., and Durham, D. K. (Eds.). Media and Cultural Studies: Keyworks. Malden, MA: Blackwell Publishing.
Mumford, L. (1934). Technics and Civilization. New York: Harcourt, Brace.
________. (1970). The Myth of the Machine: The Pentagon of Power. New York: Columbia University Press.
Naofusa, H. (1983). Traditional Cultures and Modernization: Several Problems in the Case of Japan. Kokugakuin University. Retrieved June 9, 2009, from www2.kokugakuin.ac.jp/ijcc/wp/cimac/hirai.html.
Nye, D. E. (1990). Electrifying America: Social Meanings of a New Technology, 1880-1940. Cambridge: MIT Press.
O'Neill, M. (1993). The Roar of the Crowd: How Television and People Power are Changing the world. New York: Times Books.
Oakes, T. (1998). Tourism and Modernity in China. London and New York: Routledge.
Olds, K., and Yueng, H. W. (2004). Pathways to Global City Formation: A View from the Developmental City-state of Singapore. Review of International Political Economy 11 (3), 489-521.
Ong, A. (1999). Flexible Citizenship: The Cultural Logics of Transnationality. Durham, NC: Duke University Press.
Ong, W. J. (2002). Orality and Literacy: The Technologizing of the Word. New York: Routledge.
Onishi, N. (2006). Arising Korean Wave: If Seoul Sells It, China Craves It. Retrieved June, 19, 2009, from www.iht.com. www.iht.com/articles/2006/01/02/news/korea.php.
Owen, B. M. (1999). The Internet Challenge to Television. Cambridge, Massachusetts/London, England: Harvard University Press.
Park, M., and Curran, J. (2000). De-Westernizing Media Studies. New York: Routledge.
Peters, J. D. (1997). Seeing Bifocally: Media, Place, and Culture. In A. Gupta and J. Ferguson (Eds.). Culture, Place, and Power: Essays in Critical Anthropology (pp.75-92). Durham: Duke University Press.
———. (1989). Satan and Savior: Mass Communication in Progressive Thought. Critical Studies in Mass Communication, 6, 247-263.

______. (2005). Technology and Ideology: The Case of the Telegraph Revisited. In J. Packer and C. Robertson (Eds.). Thinking with James Carey: Essays on Communication, Transportation, History (pp. 137-56). New York: Peter Lang.
Pieterse, N. J. (1995). Globalization as Hybridization. In S. L. M. Feather and R. Robertson (Eds.). Global Modernities (pp. 45-68). Thousand Oaks, CA: Sage.
Radway, J. (1984). Reading the Romance: Women, Patriarchy and Popular Literature. Chapel Hill, NC: UNC Press.
Rajagopal, Arvind. (2006). Imperceptible Perceptions in Our Technological Modernity. In Chun W. and Keenan T (Eds.). New Media, Old Media: A History and Theory Reader (pp. 277-286). New York: Routledge.
Ran, H. (1998). Dianshi Chuanbo Yu Dianshi Wenhua [TV Broadcasting and TV Culture]. Wuhan, China: Wuhan University Press.
Rantanen, T. (2005). The Media and Globalization. London: Sage Publication.
Reeves, G. (1993). Communication and the "Third World." London: Routledge.
Robins, K., and Webster, F. (1999). Times of the Technoculture: From the Information Society to the Virtual Life. London and New York: Routledge.
Rogers, R. A. (1994). Nature and the Crisis of Modernity: a Critique of Contemporary Discourse on Managing the Earth. Montreal/New York: Black Rose Books.
Rosen, S. (2003). Chinese Media and Youth: Attitudes Toward Nationalism and Internationalism. In C.C. Lee (Ed.). Chinese Media, Global Contexts. London and New York: RoutledgeCurzon.
Schivelbusch, W. (1986). The Railway Journey: The Industrialization of Time and Space in the 19th Century. Berkeley and Los Angeles: The University of California Press.
Sconce, J. (2000). Haunted Media: Electronic Presence from Telegraphy to Television. Durham, NC: Duke University Press.
Scott, A. J. (2000). The Cultural Economy of Cities. London: Sage Publications.
Silverstone, R. (1994). Television and Everyday Life. New York: Routlege.
Song, G. (2004). The Fragile Scholar: Power and Masculinity in Chinese Culture. Hong Kong: Hong Kong University Press.
Sorensen, C., and Kim, S. (2004). Filial Piety in Contemporary Urban Southeast Korea: Practices and Discourses. In C. Ikels (Ed.). Filial piety: Practice and Discourse in Contemporary East Asia (pp. 153-218). Stanford: Stanford University Press.
Spigel, L., and Olsson, J. (Eds.). (2004). Television after TV: Essays on a Medium in Transition. Durham & London: Duke University Press.
Spigel, L. (1992). Make Room for TV: Television and the Family Ideal in Postwar America. Chicago: University of Chicago Press.
Sreberny-Mohammadi, A. (1997). The Many Cultural Faces of Imperialism. In P. Golding and P. Harris. (Eds.). Beyond Cultural Imperialism: Globalization, Communication and the New International Order (pp.49-68). Thousands Oaks, CA: Sage.

Starrs, R. (2001). Asian Nationalism in an Age of Globalization. Richmond, Surrey: Japan Library (Curzon Press).

Stearns, N. P. (2006). Gender in World History. New York: Routledge.

Stephens, M. (1998). The Future of the Internet: Which Communication Revolution Is It Anyway? Journalism Quarterly, Spring.

Striphas, G. T. (2009). The Late Age of Print: Everyday Book Culture from Consumerism to Control. New York: Columbia University Press.

Sturken, M., and Cartwright, L. (2001). Practices of Looking: An Introduction to Visual Culture. New York: Oxford University Press.

Sung, Kyu-Taik. (1998). An Exploration of Actions of Filial Piety. Journal of Aging Studies, 12 (4), 369-86.

Tao, J., Zheng, B., and Mow, L. S. (2004). Holding up Half the Sky: Chinese Women Past, Present, and Future. New York: Feminist Press at the City University of New York.

Terranova, T. (2004). Network Culture: Politics for the Information Age. London and Ann Arbor, MI: Pluto Press.

Thakur, R. (1997). Rewriting Gender: Reading Contemporary Chinese Women. London; Atlantic Highlands, N.J.: Zed Books.

The Current Situation of the Cable TV, Law and Legislation in Mainland China. (1994). Taipei: Association of Mainland China Studies.

Thomas, A., and Dyson, S. (2004). Asia Pacific TV: Incorporating CASBAA Asia Cable and Satellite Guide (8th ed.) London: Informa Media Group.

Thomas, J. A. (2001). Reconfiguring Modernity: Concepts of Nature in Japanese Political Ideology. Berkeley: University of California Press.

Thompson, A. E. (2002). The Soundscape of Modernity: Architectural Acoustics and the Culture of Listening in America, 1900-1933. Cambridge, MA: MIT Press.

Thompson, B. J. (1995). The Media and Modernity: A Social Theory of the Media. Stanford: Stanford University Press.

To, Y., and Lau, T. (1995). Global Export of Hong Kong Television: Television Broadcasts Limited. Journal of Communication 5 (2), 108-21.

Tomlinson, J. (1991). Cultural Imperialism: A Critical Introduction. London/New York: Continuum.

______. (1999). Globalization and Culture. Chicago: University of Chicago Press.

Tracey, M. (1988). Popular Culture and the Economics of Global Television. Intermedia 16 (2), 9-25.

Tu, W. (1994). The Living Tree: The Changing Meaning of Being Chinese Today. Stanford, CA: Stanford University Press.

Ussher, J. (1997). Fantasies of Femininity: Reframing the Boundaries of Sex. New Brunswick, NJ: Rutgers.

Uricchio, W. (2004). Television's Next Generation: Technology/Interface Culture/Flow. In L. Spigel and J. Olsson (Eds.). Television after TV: Essays on a Medium in Transition (pp. 163-182). Durham and London: Duke University Press.

Wang, B. (2004). Illuminations from the Past: Trauma, Memory, and History in Modern China. Stanford: Stanford University Press.

Wang, J. (2000). Foreign Advertising in China: Becoming Global, Becoming Local. Ames: Iowa State University Press.

Wang, M., Yu, X., and McLean, G. (Ed.) (1997). Chinese Cultural Traditions and Modernization. The Council for Research in Values and Philosophy. Cultural Heritage and Contemporary Change. Series III, Asia, 10.

Wang, Y. (2006, July 20). Chen Duo and Talking about the Changjiang River [Chen Duo he Huashuo Changjiang]. Retrieved July 5, 2009, from Renmin Wang (www.people.com.cn), media.people.com.cn/BIG5/40606/4611678.html.

Weller, R. P. (2006). Discovering Nature: Globalization and Environmental Culture in China and Taiwan. New York: Cambridge University Press.

Wen, J. J., and Tisdell, C. A. (2001). Tourism and China's Development. New Jersey: World Scientific Publishing Company. CA: World Scientific.

Whyte, K. M., and Parish, W. L. (1984). Urban Life in Contemporary China. Berkeley: University of Chicago Press.

Whyte, K. M. (2004). Filial Obligations in Chinese Families: Paradoxes of Modernization. In Ikels, C. (Ed.). Filial Piety: Practice and Discourse in Contemporary East Asia (pp. 106-127). Stanford: Stanford University Press.

Williams, R. (1980). Problems in Materialism and Culture: Selected Essays. London: Verso.

______. (2005). Television: Technology and Cultural Form. New York: Routledge.

Williams, R., and Higgins, J. (2001). The Raymond Williams Reader. Oxford; Malden, MA: Blackwell Publishers.

Williamson, J. (1978). Decoding Advertisements: Ideology and Meaning in Advertising. London: Boyars. Distributed by Calder and Boyars.

Winner, L. (1993). Upon Opening the Black Box and Finding It Empty: Social Constructivism and the Philosophy of Technology. Science, Technology, and Human Values 18 (3), 362-78.

______. (1986). The Whale and the Reactor: A Search for Limits in an Age of High Technology. Chicago and London: The University of Chicago Press.

Wittfogel, K. (1957). Chinese Society: A Historical Survey. The Journal of Asian Studies, 16 (3), 343-364.

———. (1970). Agriculture: A Key to the Understanding of Chinese Society, Past and Present. Canberra: Australian National University Press.

Wittig, M. (1992). One Is Not Born a Woman. In M. Wittig (Ed.). The Straight Mind and Other Essays (pp. 9-20). Boston: Beacon Press.

Wurtzler, J. S. (2007). Electric Sounds: Technological Change and the Rise of Corporate Mass Media. New York: Columbia University Press.

Xu, J. (2006). Chinese Advertisement History [Zhongguo Guanggao Shi]. Beijing: Chinese Communication University Press, 2006.

Yang, X. (2003). The Study of the Narration of TV Plays [Dianshiju Xushi Yanjiu]. Beijing: Culture and Art Publishing House.

Yeh, E. Y., and Davis, D. W. (2007). East Asian Media Industries. London: BFI Publishing.
Yeh, W. (2007). Shanghai Splendor: Economic Sentiments and the Making of Modern China, 1843-1949. Berkeley: University of California Press.
Yeung, H. W., and Olds, K. (1999). Sigapore's Global Reach: Situating the City-state in the Global Economy. International Journal of Urban Sciences 2 (1), 24-47.
Yin, H. (2002). Meaning, Production, and Consumption: The History and Reality of Television in China. In S. H. Donald, M. Keane, and H. Yin (Eds.). Media in China: Consumption, Content, and Crisis (pp. 28-40). London: RoutledgeCurzon.
_______. (1998). Chinese Media Culture at the Transitional Period of the Two Centuries [Shiji Zuanzhe Shiqi de Zhongguo Wenhua]. Beijing: Beijing Publishing House.
Yúdice, G. (2003). The Expediency of Culture: Uses of Culture in the Global Era. Durham and London: Duke University Press.
Zeng, Q. (2002). Watching over the Spiritual Home of TV Plays [Shouwang Dianshiju de Jingshen Jiayuan]. Beijing: Beijing Broadcasting College Publishing House.
Zha, J. (1995). China Pop: How Soap Operas, Tabloids, and Bestsellers are Transforming a Culture. New York: New Press. Distributed by W.W. Norton.
Zhang, X., and Wang, C. (1987). The Appreciation of Contemporary Well-known Chinese TV Plays [Dangdai Dianshiju Mingpian Shangxi]. Fuzhou: Haixia Wenyi Publishing House.
Zhang, Yingjin. (2004). National Cinema. New York: Routledge.
_______. (1996). The City in Modern Chinese Literature and Film: Configurations of Space, Time, and Gender. Stanford, CA: Stanford University Press.
Zhao, C., and Zheng, M. (2006). The Analysis on Why Korean TV Series' are Popular in China. Retrieved March 20, 2009, from CCTV.com.
Zhao, Y. (1998). Media, Market, and Democracy in China: Between the Party Line and the Bottom Line. Urbana: University of Illinois Press.
Zhao, Z. (1995). The Capriccio on the Past Years [Suiyue Suixiang]. Shanghai: Shanghai People's Publishing House.
Zhong, Xueping. (2000). Masculinity Besieged?: Issues of Modernity and Male Subjectivity in Chinese Literature of the Late Twentieth Century. Durham and London: Duke University Press.
Zhu, Y. (2008). Television in Post-reform China: Serial Dramas, Confucian Leadership and the Global Television Market. London and New York: Routledge.
Zweig, D. (2002). Internationalizing China: Domestic Interests and Global Linkages. Ithaca and London: Cornell University Press.

Chinese Magazines and Newspapers

Ba Xiaoshi Yi Wai (Life Out of 8 Hours, 八小时以外). Tianjian: Tianjian People's Publishing House, 1980-1990.

Banyue Tan (Bi-Monthly Discussion, 半月谈). Beijing: New China News Agency, 1980-1990.

Dazhong Dianying (Popular Cinema, 大众电影). Beijing: Chinese Cinema Artists Association, 1980-1986; Beijing: Popular Cinema Magazine Office, 1986-1990.

Dazhong Dianshi (Popular TV, 大众电视). Beijing: Chinese TV Artists Association, 1980-1990.

Renmin Huabao (Chinese Pictorial, 人民画报). Beijing: Chinese Pictorial Publishing House, 1980-1990.

Renmin Ribao (China Daily, 人民日报). Beijing: People's Daily Publishing House, 1980-1990.

Xiandai Jiating (Modern Family, 现代家庭). Shanghai: Modern Family Magazine Office, 1980-1990.

Zhongguo Guangbo Dianshi Bao (Chinese Broadcasting and TV Newspaper, 中国电视报). Beijing: Chinese Central TV Station, 1981-present.

Zhongguo Guanggao (Chinese Advertisements, 中国广告). Shanghai: East Art Center, 1981-1990.

Chinese TV Dramas

Bianyi Jingcha (The Policemen in Plain Clothes, 便衣警察).
Director: Lin, Yuwei
Writer: Hai Yan (Fiction)
Producer: Beijing TV Station and Beijing TV Art Center
Release Year: 1987

Cuotuo Suiyue (Idle the Time Away, 蹉跎岁月)
Director: Cai, Xiaoqing
Writer: Ye, Xin
Producer: Chinese Central TV Station (CCTV)
Release Year: 1982

Da Jiudian (Great Hotel, 大酒店).
Director: Guo, Xinling
Writer: Qian, Shiming (screenplay)
Producer: Shanghai Television Drama Production Center
Release Year: 1988

Honglou Meng (The Dream of the Red Chamber, 红楼梦).

Director: Wang, Fulin
Writer: Cao, Xueqin (Qing Dynasty, 1616-1911)
Producer: Chinese Central TV and Chinese TV Dramas Production Center
Release Year: 1987

Huoyuan Jia (The Legendary Fork, 霍元甲).
Director: Xu, Xiaoming
Writers: Lu, Xinyan and Deng, Runliang (screenplay)
Producer: Hong Kong Rediffusion Television Limited (RTV)
Release Year: 1981

Kewang (Yearning, 渴望).
Directors: Lu, Xiaowei, Zheng, Xiaolong
Writer: Li, Xiaoming (Screenplay)
Producer: Beijing TV Art Center
Release Year: 1990

Shanghai Tan (The Bund, 上海滩).
Director: Tan, Ruiming
Writers: Chen, Lihua and Liang,Yonghua (screenplay)
Producer: Hong Kong TV and Broadcasting Co. (TVB)
Release Year: 1980

Shediao Yingxiong Zhuan (Legend of the Condor Heroes Series, 射雕英雄传).
Director: Wang, Tianlin
Writer: Jin, Yong
Producer: Hong Kong TV and Broadcasting Co. (TVB)
Release Year: 1983

Shisan Mei (The Legend of the Unknowns, 十三妹).
Directors: Pan, Wenjie and Situ, Liguang
Writers: Zhao, Zhijian and Zhang, Zhihua (screenplay)
Producer: Hong Kong TV and Broadcasting Co. (TVB)
Release Year: 1983

Xiyouji (The Journey to the West, 西游记).
Director: Yang, Jie
Writer: Wu, Cheng'en (Ming Dynasty, 1504-1582)
Producer: Chinese Central TV Station
Release Year: 1986

Xin Xing (The New Star, 新星).
Director: Li, Xin
Writer: Ke, Yunlu (Fiction) and Li, Xin (screenplay)
Producer: Taiyuan TV Station

Release year: 1986

Xue Cheng (The Snow City, 雪城).
Director: Li, Wenqi
Writer: Liang Xiaosheng (Fiction)
Producer: Heilongjiang Art Studies Center and Heilongjiang TV Station
Release Year: 1988

Yangjia Jiang (The Yang's Saga, 杨家将)
Directors: Du, Qifeng, Li, Huimin and Xiao, Jiankeng
Writers: Liang, Yongmei and Zeng, Shujuan
Producer: Hong Kong TV and Broadcasting Co. (TVB)
Release Year: 1985

Yikou Caibingzi (A Mouthful of Vegetable Pancake, 一口菜饼子)
Directors: Hu, Xu and Mei, Cun
Writer: Chen, Geng (screenplay)
Producer: Beijing TV Station
Release Year: 1958

Zaishang Hushan Xing (Tiger Hill Trail, 再上虎山行).
Director: Xu, Xiaoming
Writer: Zheng, Wenhua (screenplay)
Producer: Hong Kong Asia Television Limited (ATV)
Release Year: 1983

Zhong Gu Lou (The Bell Tower and the Drum Tower, 钟鼓楼).
Director: Lu, Xiaowei
Writer: Liu Xinwu (Fiction)
Producer: Beijing TV Station
Release Year: 1986

Chinese TV Documentaries

Huashuo Changjiang (Talking about the Changjiang River, 话说长江).
Producer: Chinese Central TV and Japanese NHK
Release Year: 1983

Huashuo Yunhe (Talking about the Grand Canal, 话说运河).
Producer: Chinese Central TV
Release Year: 1984

Sichou zhi Lu (The Silk Road, 丝绸之路).
Producer: Chinese Central TV and Japanese NHK
Release Year: 1979

Wang Changcheng (Looking at The Great Wall, 望长城).
Producer: Chinese Central TV and Japanese NHK
Release Year: 1991

Japanese TV Dramas

Zisan Silang (Sugata Sanshiro, 资三四郎)
Director: Kunio Watanabe
Writers: Tsuneo Tomita (novel), Kunio Watanabe
Producer: Shochiku Eiga Co.
Release Year: 1970 in Japan
Imported Year in China: 1981 by Shanghai TV

O Xin (Oshin, 阿信)
Director: Eguchi Hiroyuki, Kobayashi Heihachiro, Takemoto Minoru and others
Writer: Hashida Sugako
Producer: NHK Television
Release Year: 1983 in Japan
Imported Year in China: 1985 by Chinese Central TV

American TV Dramas

Shentan Hengte (Hunter, 神探亨特)
Creator: Frank Lupo
Producer: Stephen J. Cannell Productions
Release Year: 1984 on NBC
Imported year in China: 1988 by Shanghai TV

Daxiyang Di Lai De Ren (Man from Atlantis, 大西洋底来的人)
Director: Lee H. Katzin
Producer: Solow Production Company
Release Year: 1977 on NBC
Imported year in China: 1980 by Chinese Central TV

INDEX

About the Author

Huike Wen is an assistant professor at Willamette University, Salem, Oregon. Her research includes media history, gender and mass media, and media and society. She focuses on the cultural production in China and the interaction among different cultures in East Asia.

www.ingramcontent.com/pod-product-compliance
Lightning Source LLC
Chambersburg PA
CBHW031414220926
51612CB00004B/9
* 9 7 8 0 7 3 9 1 7 8 8 6 7 *